THE BUSINESS ROMANTIC

THE BUSINESS ROMANTIC

Give Everything, Quantify Nothing, and Create Something Greater Than Yourself

TIM LEBERECHT

An Imprint of HarperCollins*Publishers*

 For information, address HarperCollins Publishers, 195 Broadway, New York, NY 10007.

HarperCollins books may be purchased for educational, business, or sales promotional use. For information, please e-mail the Special Markets Department at SPsales@harpercollins.com.

FIRST EDITION

Designed by William Ruoto

Library of Congress Cataloging-in-Publication Data has been applied for.

ISBN 978-0-06-230251-9

15 16 17 18 19 OV/RRD 10 9 8 7 6 5 4 3 2 1

For my parents, Edith and Volker Leberecht

To romanticize the world is to make us aware of its magic, mystery, and wonder; it is to educate the senses to see the ordinary as extraordinary, the familiar as strange, the mundane as sacred, the finite as infinite.

—NOVALIS

CONTENTS

INTRODUCTION

The Flame

The flame must never go out.

That's what our instructors had told us again and again during our training. The protocol, logistics, and etiquette we were taught to follow were akin to a presidential campaign. All staff members traveling with the flame were required to be on time, on message, on values, and on top of their game, all the time. These were the Olympics, after all, and only if we showed our best selves would the Games succeed in living up to their ideal of building a better world through sports. Not only what we did, but also *how* we did it, mattered a great deal: show grace under pressure; stick to the rules, but, more important, follow the Olympic spirit; shave every morning; drink no alcohol; and always be punctual. This was not a job to be done on autopilot, and it would tolerate no egos. The flame was the star. The Olympic values (of "excellence, friendship, and respect") were our product.

I was offered this six-week gig right out of grad school through a chain of coincidences. I accepted for the thrill of it, for a chance at an unforgettable experience (which in my case included the little extra excitement of giving a speech entirely in French at the Montreal city hall, read from a phonetic cheat sheet that my local PR colleague had prepared for me). My pay was nominal, and I flew coach class around-the-world-in-six-weeks, but I didn't care about any of these aspects of the work. None of us did. We were in it for the sake of doing it, and the pride we all felt was infectious. It was the best job I had ever had.

Before coming home to Athens, the birthplace of the original Olympics and the location of the 2004 Summer Games, the flame would touch down in thirty-two former Olympic host cities, including Tokyo, Los Angeles, Montreal, Paris, London, Munich, and Moscow. The event, the Olympic Torch Relay, was meant to galvanize Olympic fans, creating an exciting prelude to the kickoff of the Games. But that year, it was even more remarkable. The flame would touch African soil for the first time in history, passing through Cairo, a city playing its first role in the Olympic story. As one of the advance press chiefs for the relay, I was up at the crack of dawn, waiting at the Cairo airport to meet the chartered 747 jet carrying the flame. But I wasn't just meeting the flame; I was meeting its entire entourage, including two around-the-clock security guards.

When the flame was finally carried off the plane in Cairo, a crowd of officials, staff, and VIPs greeted it. A special bus, escorted by a motorcade with heavy security detail and tailgated by a throng of reporters, brought it to the city center, where the first torch runners had been awaiting their turn for hours. Each one of these runners had applied in advance for their five minutes of fame with the flame, everyone following a meticulous schedule that—until the Cairo leg—had been flawlessly executed. One hour after the flame's arrival, we all stood in the middle of Tahrir Square, where tens of thousands of joyful Egyptians were chanting and celebrating, the line between order and chaos increasingly tenuous. I was meeting strangers, talking to strangers, directing strangers, all of us connected through the power of the Olympic flame. While I focused on the logistics of my job—herding reporters and staff and vans, from mile to mile, and viewpoint to viewpoint—I could feel the flame finding its path through the crowd, warming the hearts of the people of Cairo. My meals had consisted of energy bars and bottles of water, and sweat drenched my forehead and neck, and yet despite all this, I felt a hum of euphoria; I was in it and of it, fully alive.

Finally, the whole caravan and its entourage came to a halt somewhere on the outskirts of the megalopolis, with the pyramids looming as backdrop. The usual end-of-day celebration began, an official ceremony starring city officials (often the mayor), representatives of the Athens Organizing Committee, and local celebrities.

As the program started, I could see some of my colleagues drifting away, slowly fading from sight in the encroaching darkness, melding with the pyramids and the hazy contours of the desert. Under the sheltering sunset, I tried to listen to the speeches, but the spectacle took on a surreal air: a mirage of illuminated faces moving in feverish, erratic rhythms, uttering incomprehensible syllables. There I stood, sand in my shoes and in my mouth, surrounded by honking cars, the ancient monuments, the heat, the smell of gasoline, and Egyptian police; a German in Cairo representing the Athens games; working for a U.S. company from Denver that was producing the relay—all cast against the omnipresent logos of Coca-Cola and Samsung, the two main sponsors.

With the exception of the soccer World Cup, no other sports event holds such universal appeal—and no other sports event has become so commercialized. For that reason, the Olympics are often accused of moral bankruptcy and regarded with ambivalence, but in that very moment, under the Egyptian sky, the Olympic idea was as alive and pure as ever. There was something sacred even in its most profane moments. Niels Bohr, the legendary

Danish quantum mechanics physicist, once stated that the sign of a great truth is that its opposite is also a great truth;[1] no statement better describes the contradictions at the core of the Olympics.[2] The games are business-as-usual, but also, still, the most romantic idea and endeavor imaginable. Political mine fields, hard work, dirt and dust and sweat, brand guidelines, ambush marketing, airtime, swag, talking points, and spreadsheets, overpriced and oversold—yes, all of that, and yet so much more. In the flame and in the faces of the Egyptians I met, I saw the Olympic spirit: the idea that peaceful competition—the Latin root for competing, *competere*, means "to strive together"—allows us to unleash our full human potential and to connect as citizens of the world.

The Olympics are what you see in them; they are what you make them, what you want them to be. Like the greatest of experiences, their very inadequacies, their imperfection, is what lends them their romance. They leave space for our imagination. They leave us wanting more.

Money and meaning, commerce and culture, transaction and transcendence: I have always been drawn to these tensions. My grandfather was a filmmaker, and my father is the managing director of an executive education firm. When I was six years old, I created my own imaginary corporation, and at the age of twenty-one, I founded my own music publishing company (the former was far more successful). In my early twenties, I played in a band and released two albums, a process that taught me more about collaboration than anything else I've done in the whole of my career. I studied both liberal arts and business administration. As a student, I read German philosophers; as a businessman, I read the *Wall Street Journal*. I had always wanted to become an artist—and ended up becoming the "marketing guy."

In business, I have found the same beauty and intensity that I experienced singing my heart out in a concert. Today I no longer play in a band or even think of myself as a musician; business is my stage. I no longer write songs; I write e-mails, memos, articles, presentations, and strategic plans. In client pitches and town halls, at conferences and networking events, in team brainstorming sessions or sitting alone at my desk, many of my most transcendent moments now occur through work.

As a marketing executive, I view business as one of the greatest adventures of the human enterprise—if not *the* greatest. But I am not just a businessman: I am also an unapologetic romantic. I believe that the world would be a better place if we had more romance in our lives. I believe that promise trumps fulfillment. I believe that emotion eats reason for breakfast. I am not a daydreamer, idealist, or social activist. I am a Business Romantic.

Throughout my career—as chief marketing officer of a product design and strategy firm (Frog Design), an IT outsourcing firm (Aricent), or most recently an architecture and design firm (NBBJ)—I have come to appreciate markets as a powerful vehicle for making connections, for creating value beyond merely facilitating transactions. In the most tangible of our exchanges—at the farmers market, say, or in an online marketplace like Etsy—we discuss our needs and wants with people who are equally invested in them. We converse. We find common cause. We use the markets to communicate. No one has put it better than philosopher Robert C. Solomon:

> Market systems are justified not because of efficiencies and profits, but because humans are first and foremost social and emotional beings, and markets provide a sympathetic community for social exchange.[3]

When we go to market, we present ourselves to the world: it is the beginning of business, the beginning of romance.

You may say: "Wait a minute. This is naive. You are romanticizing business. In today's gloomy economy and partisan politics, it is almost irresponsible to have such a romantic view of business. How can we be romantic in such a depressing environment? The chasm between market and meaning is simply too wide to cross."

Is it, though? Should it be?

Why must business exist within an entirely transactional framework?

What if the process is as important as the end result?

Can we maintain profit margins while also making experiences that put us in touch with the wondrous, the delightful, and the mysterious?

Isn't it time to bring our fullest selves to the job?

What if we could find romance in and through business?

"Business" and "romance": these two words touch on the conflict within all of us about the role that business can and should play in our lives. When we discuss human-scale markets, this unlikely pairing can feel intuitive and comforting; but when markets grow more abstract, when they begin to look more like systems, and the humans inside them scale up to form a workforce, or an industry, the words "business" and "romance" start to sound suspicious.

And yet, if we believe that the benefits of the market economy generally outweigh its faults, we need not attempt to completely contain or resolve this conflict. With the exception of government, few cultures have a greater impact on us than business: as employees, consumers, and even citizens. We live in a market society, whether we like it or not. What we buy and what we do for a living reflects (and even determines) who we are. Our career paths offer up many of our most salient opportunities for self-realization, and most of us spend the majority of our lives at work. For many of us, our coworkers are more intimately involved in our lives than our neighbors or friends, or even our families. In fact, studies suggest that we are likely to have more friendships than any other kind of relationship in the workplace (like boss-subordinate or mentor-protégé relationships).[4] Such "blended relationships"—bringing work into the home and home into work—are one of the hallmarks of our connected age.

In all these ways, business binds us in its various chains of meaning. So when it comes to bringing more romance to our lives, why not start here?

When I ask friends and colleagues how they feel about work, I hear this:

- *"I am a creature of the office; I thrive in this fixed environment."*
- *"At work, I get to make my jokes and wander the halls, and generally do all the socializing that I would never think to do in my neighborhood or walking down the street."*
- *"I always have the freedom to work from home, but why would I? The office is where I belong."*

When we seek meaning, work is our arena. David Whyte, a poet and business consultant, perfectly captures the essence of this idea: "Work is difficulty and drama, a high-stakes game in which our identity, our self-esteem, and our ability to provide are mixed inside us in volatile, sometimes explosive ways," he writes. "Work is where we can make ourselves; work is where we can break ourselves."[5]

This is precisely why so many of us also suffer greatly in business. We suffer under the constrictions of the traditional market system and models of decision making that assume we are fully rational beings. And we suffer when we continue to create false distinctions between our business personas and other, larger parts of our humanity, when we divorce business from our emotional, intellectual, and spiritual needs.

Many of us clamor for more. We are in business, as consumers, employees, and entrepreneurs, because we love business. We love the drive of it; we love its opportunities for connection and social exchange. Some of us start our own businesses; others work at the forefront of innovation or management. Still others work in creative fields such as music and publishing—industries ever on a tightrope between commerce and culture—while too many of us still speak softly, if at all. We stash our wistful longings away when we enter our cubicles in the morning, our longings for an opportunity to express our truest selves at work, for an experience that makes us feel fully alive in our jobs, throughout our careers. This book is speaking to all of us—to each and every one of us who feels that business-as-usual is disenchanting all that is magical and meaningful about our daily experiences as professionals or consumers.

After reading this book, you will have more ways to open yourself up to the delights, the mysteries, the moments of transcendence, and even the hard-won sorrows of an everyday life in business. You will shimmer and glow amid the doom and gloom. You will better understand how to design spaces for these experiences, both with your colleagues and for your customers. You will walk away with Rules of Enchantment that show you how to act on your romantic beliefs and how to articulate them to others. And, surprisingly enough, this shift in perspective and attitude will open up unforeseen opportunities all around you, with the most unexpected of partners. You will begin the romance anew.

Although this book does not present a new management paradigm or economic theory, it does intend to upend conventions. It is a clarion call for all those for whom mere business excellence and efficiency are not good enough. In this way, it is a small book, human-sized and intended for an individual's hands. It encourages you to have a romantic view of business: to act differently, but, first and foremost, to see, feel, and be different. It begins at the personal level, but ultimately this shift has the potential to instigate wide-scale institutional and systemic change. We may not play a different game, but if we play by different rules, we will end up with a better one.

In the following chapter, "The New Desire for Romance," I look more closely at the current zeitgeist. Why do we need more romance, and how does it manifest itself in our first loves and first jobs? Next, in "Meet the Business Romantics," I introduce five individuals, a couple, and a family who have found and sustained this romance in their relationship to business. You will learn to recognize the romantic characteristics both in other people as well as in yourself.

These chapters are followed by the "Rules of Enchantment." How can we use small acts of significance and rituals to make our lives as workers and consumers more meaningful? Where do we cultivate romantic experiences of friction, conflict, mystery, and ambiguity? How can we bring some of the "art of life" back into our everyday business? And how do we fall in love with our job again and again?

By the time you reach the final chapter, "The New Romantic Age," you will be well positioned to keep your own flame alight while also igniting that of others. And—romance isn't easy—I will discuss some of the caveats and conundrums, the challenges that come with scaling Business Romance. Moreover, you will learn more about the Business Romantic's role in society and how you can promote change that transcends the individual life and ushers in a new romantic age. In addition, the Appendix—"The Business Romantic Starter Kit"—will offer you practical resources, tips, and instructions.

By writing this book, I hope to instigate the quietest of revolutions. It is time for us to stand together. It is time for us to speak out. Who are *we*? We are the men and women of business who are ready for more. We are the Business Romantics.

PART ONE

Kindling

CHAPTER 1

The New Desire for Romance

MARKET SOCIETY IN CONFUSION

According to a 2013 Gallup poll conducted in 140 countries, only 13 percent of employees worldwide are fully involved in and enthusiastic about their jobs. Sixty-three percent are "not engaged" and "lack motivation." About 24 percent are "actively disengaged," meaning "they are unhappy and unproductive at work, and liable to spread negativity to coworkers."[1]

Things look even more grim for leadership in business: Edelman's 2013 Trust Barometer reveals that academics, technical experts, and members of midlevel management are nearly twice as trusted as chief executives.[2] In its Outlook on the Global Agenda 2014, a survey of more than 1,500 leaders from the public and private sectors, the World Economic Forum identified "a lack of values in leadership," "diminishing confidence in economic policies," and "widening income disparities" as three major trends affecting societies worldwide.[3]

In the same vein, the Organisation for Economic Co-operation and Development (OECD) reports that social inequality in most industrialized nations has grown significantly since the global economic crisis broke out in 2008, with the wealthiest 10 percent of the population increasing their wealth 9.5 times as much as the poorest 10 percent in 2010.[4] In his much discussed book, *Capital in the 21st Century*, French economist Thomas Piketty contends that we have regressed to a "patrimonial capitalism" that resembles nineteenth-century levels of wealth distribution, concentrated among family dynasties.[5] In the United States, the Occupy Wall Street movement has helped us frame the growing income gap as the difference between the 99 and 1 percent. But research shows that this relationship is even more tilted toward the super-rich, the 0.1 percent. While the top 1 percent of American households take in about 22 percent of income (including capital gains), 0.1 percent of households own one-fifth of the country's wealth.[6] A 2013 report by the Aspen Institute concludes we may be shifting to a "Power-Curve Society" in which prosperity no longer follows a bell-curve distribution but instead accumulates at the top strata of winner-take-all societies.[7]

The innovations of the digital economy accentuate this trend. It is not only the future that is unevenly distributed, to paraphrase William Gibson's famous quote—so is the creation of value. When Facebook bought WhatsApp for $19 billion in 2013, it paid a price equal to $345 million for each of the fifty-five employees.[8] In his book *Who Owns the Future?* Jaron Lanier points to the implications for the workforce: "Kodak employed 140,000 people, Instagram 13."[9] We have entered a binary age, where software not only "eats the world," as Silicon Valley venture capitalist Marc Andreessen proclaimed,[10] but apparently also the middle class. Every day brings another newspaper report on social inequality, whether it is the disappearance of the market for middle-class consumer goods[11] or President Obama calling income disparity the "defining challenge of our time."[12] Wall Street has taken note: although investment bankers still receive exorbitant bonuses, they have lost much of their masters-of-the-universe appeal. In fact, the current environment has brought forth entirely new types of Wall Street characters. Suddenly the media is rife with confessionals such as "Why I Am Leaving Goldman Sachs"[13] and stories of former financiers seeking redemption for money addiction by working in the not-for-profit world.[14]

But we don't need the media to show us what is right in front of our faces every day. The most powerful proof of confusion and disenchantment is the mood of the culture all around us: on the streets, in our subways, at our universities and schools, and in our very own offices. We survived the financial crisis, and we've now landed in the brave new world of the current recovery: between ever-changing arguments about quantitative easing, the slowing of growth in China, and the emergence of possible technology and real-estate bubbles, is it any wonder that we have a sense of whiplash about traditional business and economic models? Many of us feel that we are giving more and more for less and less in return. We are working longer hours (in fact, Americans work approximately eight weeks longer per year than in 1969, but for roughly the same, inflation-adjusted income[15]), accumulating greater debt, and watching middle-class promises such as home ownership and a solid education for our children erode. Millennials, the age group between eighteen and thirty-three, have higher levels of student loan debt, unemployment, and poverty, and lower levels of wealth and personal income than their two immediate predecessor generations (Gen Xers and Boomers) had at the same stage of their life cycles.[16] For the first time on record, a generation is economically regressing, rather than progressing.

In Silicon Valley—a culture indoctrinated in the edicts of tech-optimism—new forms of social technology promise to close the gaps left by diminished governments, civic structures, and media organizations still struggling to find their bearings in the wake of the 2008 crisis. But do we really think software companies such as Amazon, Facebook, and Google, and all of their younger offspring can address our most important social and ethical questions? A growing cadre of cultural commentators and philosophers is voicing concerns about supplanting the responsibilities of citizenship with a myopic belief in technology. Evgeny Morozov, one of the most prolific and sharp-tongued in this group of critics, scoffs at such "solutionism," describing it as "an intellectual pathology" that recognizes problems "based on just one criterion: whether they are 'solvable' with a nice and clean technological solution at our disposal."[17]

Moreover, the speed of technological advances is outpacing our institutions and moral capacities. We live in an age where we make science-fiction movies to catch up with reality. On any given day, we need to develop cogent opinions on developments ranging from surveillance—by both companies and our own government—to cyberwarfare, bioterrorism, genetic engineering, and political uprisings amplified by tools of social media. And that's just the morning news! Before we can fully process the pace of change, much less develop a moral and ethical perspective on it, innovations on the cusp of science and technology have already begun to transform us as individuals and as societies.

As these technological innovations create economic disruptions, they also destabilize our value systems. Pope Francis, *Time*'s "Person of the Year 2013," lashed out against the "tyranny of unfettered capitalism," denouncing "trickle-down" economic policies, financial greed, and consumerism. The head of the Catholic Church made it clear that we are in grave danger of losing our moorings, bowing only before the gods of money.[18] A business mindset has infiltrated the most private aspects of our lives, bit by bit, tit for tat. We out-schedule each other to assure ourselves of our success, and the only ecstasy we can still find lies in a state of constant overwhelm.[19]

Even our friendships are starting to suffer. A group called Lifeboat, which describes itself as a movement "celebrating deep friendships," recently conducted a full-scale study on the state of friendship in the United States.[20] The report was the first of its kind, and its results were sobering. Only a quarter of adults are truly satisfied with their friendships. Despite the rise of social networking sites and increasing opportunities to connect online, most Americans told Lifeboat that they would rather have fewer, more meaningful connections than a greater quantity of friends. According to the study, friendship in the United States is in a state of "crisis."

This sense of isolation is playing out like a house of mirrors against our developing digital landscapes. Facebook and other social media tools were designed to make us feel more connected, but our current anxieties of income inequality and work dissatisfaction are only amplified in the digital commons. There is always someone else in our network having more fun, making more money, or getting more psychic value out of those connections. Millennials retreat to individualism and a growing detachment from institutions such as religion, marriage, or political parties. A 2014 Pew survey[21] showed that the social capital of these "digital natives" is mainly generated through social media networks. The need for self-expression and connection is high (55 percent of Millennials have posted a "selfie" online), but trust in others is low: just 19 percent of them say most people can be trusted, compared with 31 percent of Gen Xers and 40 percent of Boomers.

Against this backdrop of confusion and insecurity, Millennials are looking for a greater sense of meaning and community through work. But they are not the only ones. In an op-ed piece for the *New York Times*, the economist Jeffrey Sachs coined this moment in time—following the overleveraged glut of our most recent Gilded Age—"New Progressivism."[22] As a result of these shifts in the zeitgeist, alternative forms of doing business are popping up everywhere. Companies are getting certified as "B Corps," for-profit corporations with a mandate to solve social and environmental problems;[23] nonprofit organizations such as Ashoka are teaming up with Fortune 500 corporations to promote Corporate Change Makers;[24] the Maker Movement is finding its greatest source of institutional support from the big-box stores;[25] capitalist visionaries such as Sir Richard Branson and the former CEO of sportswear maker Puma, Jochen Zeitz, are launching the B Team, a coalition aiming to connect business leaders in delivering on social and environmental goals;[26] Whole Foods co-founder and CEO John Mackey is evangelizing the idea of "Conscious Capitalism;"[27] and journalist and consultant Tony Schwartz's Energy Project is promoting more purposeful employee engagement based on the concept of an individual's "energy."[28]

What is happening?

In 2001, Aaron Hurst, a serial social innovator who started his first venture at the University of Michigan at the ripe old age of sixteen, founded the Taproot Foundation. The idea was simple but revolutionary for its time: linking business professionals with opportunities to do pro bono work. During his many years with Taproot, Hurst received more than twenty-five thousand letters written from business practitioners all over the world. They all described to him what they hoped to gain from a pro bono experience. Why in the world would otherwise successful business professionals seek out opportunities to give away their time for free?

Of course, if we posed such a question within a civic or religious sphere, it would sound absurd. Service to others and a desire for deeper connections are in the very DNA of these domains. In business, however, we are taught to think of ourselves as machines: agents of optimization, efficiency, and productivity. Hurst, on the other hand, identified four distinctly human drivers behind the strong demand for pro bono opportunities: (1) meeting new people; (2) craft building; (3) finding problems worth solving; and (4) connecting with people in the community.

He went on to shape his argument into a book and platform he calls *The Purpose Economy*[29] and subsequently merged with social design firm Imperative to promote "purposeful business." The Purpose Economy, Hurst argues, is the next paradigm after the decline of the Information Economy. Purpose, in his words, is "empowering people to have rich and fulfilling careers and lives by creating meaningful value for themselves and others." This may seem like a radical transition—moving away from information into something even more vague and, arguably, more esoteric—but Hurst is certain that purpose will be the major driver for the next generation.

His ideas, of course, are indebted to the management guru Peter Drucker, who, decades ago, established the concept of "purpose-driven business," arguing that companies needed a raison d'être, a higher mission, to outlast the competition. But the ideas are also completely of the moment.[30] Shifts in demographics—more women in the workplace, more immigrants, and, most important of all, the presence of Millennials, which will represent 50 percent of the U.S. workforce by 2020[31]—are creating a more progressive labor force. As Hurst said, "Gen X has always been invested in social values, but, for Gen Y, it is an imperative."

Michael Norton, a professor of management at Harvard Business School and a coauthor of the book *Happy Money*,[32] told me: "The job of human resources used to be bonuses and raises, hiring and firing and selecting new employees. Today, human resources has been tasked with 'wellness' or 'How do we make our employees happier?' People are happy with more money, but it's less clear that hard economic incentives are really the way to go."

Norton and his colleagues are currently working out experiments that investigate some of these alternative incentives. In one of the experiments, employees at an Australian bank, given an opportunity to donate money to a charity, reported significantly more satisfaction and happiness at work. In another experiment, employees at a pharmaceutical firm in Belgium performed better after gifting their fellow teammates money. These, and other, "pro-social" incentives are proving more effective than money as a driver of productivity. Research by the *Harvard Business Review* and the Energy Project suggests that employees who find meaning in their work are more likely to stay with their organizations, experience higher job satisfaction, and are more engaged.[33] According to a 2014 survey by Deloitte, executives and employees who work full-time for an organization with a strong sense of purpose report much higher confidence in their organization's competitiveness and growth prospects than those lacking a sense of purpose.[34]

Kyla Fullenwider, however, a social designer and cofounder of Imperative, told me that justifying purpose, happiness, and other alternative metrics with a business case for greater productivity and higher growth feels "antiquated": "The real innovators and the early adopters don't even use the term 'business case' anymore," she said. "It's really much simpler than that. We're trying to make employees happier because we're all people. The justification is obvious. Or it should be. That's the goal."

For enlightened economists, consequently, happiness is no longer a means to an end; it is the end. The United Nations issued a Happiness Resolution and launched an International Happiness Day;[35] the *Harvard Business Review* featured "the happiness factor" on its cover in 2012,[36] and online retailer Zappos, among other firms, implemented an "employee happiness index" inspired by the Kingdom of Bhutan's Gross National Happiness Index,[37] even spinning off a whole platform for "Delivering Happiness," including a "Chief Happiness Officer."[38]

But the conversation transcends happiness and points to something even loftier: meaning. Meaning is a notably different desire from the pursuit of happiness, as research by Roy F. Baumeister and others has shown.[39] One can lead a happy life devoid of mean-

ing, and one can hav
happy. Happiness is li
is profound, commur
almost always implies
nity, whether real or

Meaning has ma
Technology worker
promote the concep
of the World Econo
meditation sessions
network Wisdom 2
insights on the mind, and
incorporate spiritual insight, meditation, emotional intelligence, and other "food for the soul" techniques into the corporate agenda. Companies such as Aetna, General Mills, Nike, and Target start their working days with meditation or yoga techniques, and Google offers its employees a "Search Inside Yourself" class that is said to sell out instantly. A new generation of leaders explore the "third metric"—alternative ways to measure a successful, or better, more meaningful life, defined by giving back, wonder, well-being, and wisdom.[40]

More than sixty years after the publication of *Man's Search for Meaning*,[41] Viktor Frankl's seminal book, "meaning" seems to have become the highest common denominator of an entire generation. A 2011 report commissioned by the Career Advisory Board found that "sense of meaning" was Gen Y's single most important indicator of a successful career. This desire for more meaning is associated with new optimism: in a Telefonica survey of twelve thousand Millennials in twenty-seven countries, 62 percent of respondents believed they could make a local difference, and 40 percent believed they could make a global difference.[42] While their predecessors in Gen X bemoaned the bureaucracy and corruption of corporate entities and institutions, Millennials simply

ant. Their generation embraces what fu-
once described as the "adhocracy":[43] modu-
etworked organizations that come together and
asily. The locus of power dynamically shifts across
rk instead of remaining static in one single institution
nization.

ut how exactly does romance relate to these networked con-
epts of purpose, happiness, and meaning? Can a purpose-driven life be a romantic one? Is romance a requisite for meaning? Does more romance mean more happiness? Before I go any further, we need to draw some important distinctions between the tenets of these larger cultural movements and the themes outlined in this book. Business Romantics certainly seek out greater meaning in work and agree that economic incentives are only a small part of what makes work rewarding, but their value set is fundamentally different. Whereas a purpose-driven business will only succeed if it achieves some clearly defined "good" in the world, Business Romantics find as much worth in the process as in the end product. As the adage goes: "It's about the journey, not the destination." In this way, romantics prioritize the actual experience over the institutional goal. For example, a good friend of mine works for a company built entirely around corporate social responsibility, and he shared his surprising—albeit private—frustrations with me: "Sometimes I find myself feeling demoralized. I love the world of marketing, but this company is almost punitive toward my passion. I'm being told, in subtle ways, not to sell our products. I feel like I'm in a 'do good' backwater: all the cutting-edge excitement is happening out in the more competitive spaces."

Being a romantic company is not synonymous with being purpose driven or socially responsible. To the romantic, social purpose matters, but learning, excitement, and adventure matter just as much, if not more. Romance also doesn't necessarily equate with morality; in fact, as we will see, it can sometimes reveal its dark side in moments of utmost intensity, uncertainty, conflict, and turmoil. You may work for a much-admired firm, devoted to a meaningful cause, but still feel an utter lack of romance. And you may find more romance working at Goldman Sachs than at a humanitarian organization. You can do good without feeling good.

Business Romantics certainly relate to the quest for happiness, purpose, or meaning, but they are ultimately searching for something different, something much more elusive and potentially more incendiary. A noble mission around a social cause is only one of any possible avenues that may lead to the heightened experience.

When I think back to my work with the Olympic Torch Relay in Cairo, the experience was romantic precisely because it was so conflicted, so contradictory. I felt a strong sense of purpose even as I felt the sponsors breathing down my neck. In the midst of the profane, we simply had to keep the flame alight. There was an intensity about that mission that left an indelible mark on my life. Did it make the world a measurably better place? Who knows? Was it an experience that brought everyone who was a part of it to life? Absolutely.

QUANTIFIED SELVES

At the beginning of the last century, the German sociologist and economist Max Weber used the term "great disenchantment" to portray the pervasive regime of the modern industrial society promoting a bureaucratic, intellectualized, and secularized view of the world.[44] Weber bemoaned that scientific understanding and technical rationality had formed an "iron cage" that forced spirituality into the margins of our lives: "Precisely the ultimate and most sublime values have retreated from public life either into the transcendental realm of mystic life or into the brotherliness of direct and personal human relations," Weber observed. In hindsight, his description of humanity in his seminal work *The Protestant Ethic and the Spirit of Capitalism* (1905) appears to be as somber as it was prophetic: "Narrow specialists without minds, pleasure seekers without heart; in their conceit, these nullities imagine they have climbed to a level of humanity never before attained."

More than a century later, humanity, on the ascent to yet another level, is undergoing another great disenchantment—this time propelled not by the industrialization but by the datafication of our markets, societies, workplaces, and relationships. You've probably heard the stunning stats: the human race generates as much data in two days now as it did in all of history before the year 2003, and the amount of data is doubling every two years to forty thousand exabytes (40 trillion gigabytes) by 2020. (As a point of reference, a single exabyte of storage can contain fifty thousand years' worth of DVD-quality video.) In the time it takes you to read this chapter, the human race will produce the same amount of data that currently exists in the U.S. Library of Congress.[45] Big Data indeed.

In the fall of 2012, I spent an hour with the managing director of Google's R&D center in Israel, Yossi Matias, and we launched into a conversation about algorithms and intuition. "Intuition itself is just an algorithm," the Google executive, a highly trained engineer, contended, "it's made up of millions of impressions that we input into the brain." He suggested we reconstruct and simulate this process. The goal, in his view, was simply to create better algorithms. His argumentation was cogent and disarming, but I left wanting more.

From number-crunching behemoths such as Google and Amazon, which generate giant repositories of user data to create highly targeted transactions, to the "Quantified Self" movement, spurred on by a bevy of new products and apps designed to help consumers enhance productivity, health, and fitness—these purveyors of data, big or small, promise to make our lives better. And they do. In a way. It is remarkable that we are now able to catalog exactly when we sleep best and which protein bars contribute to the fastest splits in our training runs. More important, breakthroughs happening at the forefront of fields such as personalized medicine and disaster management are already saving lives. But the fixation on data also leaves us feeling wistful, perhaps even a bit melancholy. If algorithms delve into every corner of our lives until nothing inexplicable remains, the loss will be tremendous. The faster we move from automated production to automated decision-making, the more human agency we risk giving up. The more we reduce our experiences—transcendent, fraught, delightful, frightening, whatever they may be—to a series of stark data points and engineered touch points, the more we chase the magic out of them.

When, in 2014, Malaysia Airlines flight 370 mysteriously disappeared from the radar, and passengers' loved ones and the world at large were in the dark about the plane's fate, the essayist and author Pico Iyer wrote a poignant column in which he reminded us of the "folly of knowing": "Whatever the field of our expertise, most of us realize that the more data we acquire, the less, very often, we know. The universe is not a fixed sum, in which the amount you know subtracts from the amount you don't."[46] Despite Big Data, ever more extensive surveillance, and our fervent impetus to know everything, Iyer suggests we embrace the humbling boundaries of our knowledge and leave space for the unknown: "Even if we do learn more about the fate of the airliner, it's unlikely that all of our questions will ever be answered. And the memory of how much we didn't know—and how long we didn't know it—ought to sober us as we prepare for the next sudden visitation of the inexplicable."

In business, there is little space for the inexplicable. Knowledge is often equated with exact measurement, and the overriding mantra is: "You can only manage what you can measure." Big Data has now also entered the workplace, tracking not only employee productivity but also social interactions—"social physics," as computer scientist Alex Pentland calls this new genre of sociometric data.[47] He refers, for example, to a smartphone app called "meeting mediator" that shows who is dominating the conversation in a meeting. It is no wonder that some scholars express concern over such new possibilities of employee surveillance and even frame it as "Digital Taylorism."[48]

To be sure, algorithmic measurements will provide more managerial insights, but we need look no further than the most recent financial crisis for evidence of how poorly we manage what we believe we can measure. In fact, failed mergers, failed product launches, reputational crises, and social media PR disasters—just these sorts of cultural disconnections and disruptions within organizations and between brands and their audiences—show us the importance of better managing what we *cannot* measure.

We have begun to explore alternative definitions and metrics of value creation such as happiness, purpose, and meaning, and we are now turning our analytical tools to quantifying and exploiting them. I welcome that we measure a different kind of value, but only if we don't forget to value what is *not* measurable.

There must be a place for the inexplicable alongside the explicable, and for the implicit in the midst of the explicit. In fact, the greatest leaders need to be Business Romantics at heart. F. Scott Fitzgerald put it best when he defined acute intelligence as the ability to "hold two opposing ideas in the mind at the same time and still retain the ability to function."[49] Leaders need this expansiveness of thought to synthesize the necessary messiness of our business lives—the competing realities that represent our societies' growing complexity. We must resist the temptation to reduce this messiness to mere quantitative terms.

The capacity to withstand the uncertainty of our daily existence is what allows us to maintain our most productive working relationships. As romantics, we consider human error a tool for self-discovery, and we appreciate the vagaries of the Un-Quantified Self. We embrace nuance; we appreciate intention as much as (and perhaps even more than) outcomes; we honor the inevitabilities of unpredictability—and failure. All these articles of faith resist algorithmic formulation, and yet they form the foundation for some of our most inspired acts of management. We *can* manage what we cannot measure: we do it every day.

A friend of mine, a novelist, recently participated in a residency of sorts along the coast of Panama. She was part of a select group of artists and scientists invited to join an Austrian art collector on her yacht for a week of cultural exchanges. The goal was to create a bridge of discourse and understanding between the humanities and the sciences, or what the British scientist and novelist C. P. Snow once famously described as "the Two Cultures."[50] My friend, paired with a group of engineering students from the Massachusetts Institute of Technology (MIT), described the engineers' desire to problem-solve even in conversations about ethics, identity, and culture. "They had answers before the other participants had even formulated their questions," she told me. Failure? A "pivot" on the path to ultimate success. Morality? A matter of context and better, data-driven decision making. Love? An algorithm if it works, a sentimental distraction if it doesn't. My friend was chilled by the laser-sharp rhetoric of the students, by their swagger that confused analytical smarts with intellect. For her, the experience called to mind a line by the Spanish philosopher José Ortega y Gasset: "I wish it would dawn upon engineers that, in order to be an engineer, it is not enough to be an engineer."

To me, the poor chemistry between the two groups on the yacht speaks to a deeper antagonism in society: technologists don't know what they don't know until they know it. In contrast, artists and romantics, by the very nature of their work, live with the tensions of ambiguity, conflict, doubt, and hesitancy. The humanities are our essential fortification as we defend ourselves against an entirely utilitarian engineering mind-set. They help us celebrate and keep sacred what we don't know. They guide us in confronting the most soulful of inquiries: Who are we in the face of natural forces? Who are we in the face of oppressive political regimes? Who are we in the face of our vocational callings? *What does our life's work really mean?*

At the commencement ceremony of Brandeis University in 2012, the *New Republic* literary editor Leon Wieseltier addressed the graduating class as "fellow humanists."[51] He identified encounters with great works of art—whether texts, images, or objects—as a "bulwark against the twittering acceleration of American consciousness." Culture, he announced defiantly, had become the new counterculture. The romantic tradition—its art, literature, philosophy, and history—once conceived of the self as a soul, both mercurial and inviolate. As we contemplate the slow retreat of the humanities, these soulful inquiries have shifted. Harvard University recently reported that the number of degrees in the humanities had fallen across the university—as well as from 14 percent to 7 percent across the nation between the years of 1966 and 2010.[52] While these numbers were not undisputed,[53] the ensuing debate illustrated the underlying dilemma: a crisis of confidence in the relevance of the humanities. Today many of us are drawn to experts donning lab coats; we look for scientific certification; we seek out correlations, not causes. Our culture's fixation on science has sanitized our earlier, more romantic notions of a dark and stormy spirit; the mysteries of temperament and mood have been replaced with the specifics of cells, neurons, and synapses. In the shadow of quantification, the humanities have come to be seen as honorable but inconsequential. They might capture something about our past, but they have nothing to teach us about our future.

How far we have come from the days—only decades ago—when a select group of scholars labored over the creation of the now-famous core curriculum! First at Columbia University, and then, later, at the University of Chicago, professors devoted four-year curricula to exploring the hundred to one hundred fifty books identified as the moral architecture—the great classics—of the Western World.[54] The contemporary liberal arts college was loosely designed around the rigorous study of these select texts. Close to a century later, all that has changed.

Our core curriculum has eroded—with help from the deconstructionist questions of authorship and ownership sounding out in anger across college campuses throughout the eighties and nineties. What was once the heart and soul of our education, the foundation of our most basic notions regarding our humanity, has now become a field of study pursued only by dreamers and rebels. These students, Wieseltier's counterculture, graduate with no concrete skills. And even the promise of their greater cultural understanding is regarded as dubious: the only culture that seems to capture our sustained attention is the cultural milieu of high-tech and capital-infused corridors such as Silicon Valley.

The dramatic consequences advanced digital technologies will have on our educational institutions are captured in the recent book *The Second Machine Age*, written by the MIT economists Erik Brynjolfsson and Andrew McAfee.[55] The authors argue that the exponential growth of innovation in computation is now poised to completely surpass not only our physical skills but also our cognitive ones. Has the Singularity finally arrived? Are we set to lose ourselves—and the aura of humanity itself—in this new era of the Quantified Self?

The authors—neither entirely optimistic nor pessimistic about the current upheaval in the labor market—argue for a third way, a balanced approach that acknowledges the vital role a broad education has to play: "There's never been a better time to be a worker with special skills or the right education, because these people can use technology to create and capture value. However, there's never been a worse time to be a worker with only 'ordinary' skills and abilities to offer, because computers, robots, and other digital technologies are acquiring these skills and abilities at an extraordinary rate." "Number-crunching computers will replace number-crunching managers," Tim Laseter predicts in a related article on "management in the second machine age."[56]

The romantic makes a passionate case for the foundations of our education. The humanities are our most treasured—and useless—"special skills." And it is their very uselessness—their determination to remain uncorrupted by models of efficiency and optimization—that is their saving grace. The market system, currently sucking them dry, is the very same system that will reveal their necessity.

Quantified, efficient selves—the good news is: we've been there before. The original Romantic Movement—toward the end of the eighteenth century and into the nineteenth—emerged in response to the Industrial Revolution and the Enlightenment. When the pendulum swung society to its most extreme expression of rationalism and empiricism, society—artists and philosophers, in particular—demanded that it swing back. And swing it did . . .

WILD AND WONDERFUL

It was June of 1816 on Lake Geneva, Switzerland, and bad weather was slowly starting to roll in again. One year earlier, a huge volcano erupted in Indonesia, sending a vast cloud of ash across the Northern Hemisphere. In Europe, 1816 came to be called the "Year Without a Summer,"[57] the entire season plagued by cold winds and constant downpours. Amid the usual rotation of visiting English tourists to the Lake Geneva area of the Swiss Alps, an altogether different group was settling in for a summer season of passion, transcendence, and delirium—qualities not typically associated with a society renowned for efficiency and pragmatism. The impromptu gathering of some of Western culture's most iconic "romantics" included poets Lord Byron, Percy Shelley, his mistress Mary Shelley (born Mary Wollstonecraft Goodwin), and her stepsister, the "comely" Claire Claremont. Defying "an almost perpetual rain," Mary Shelley later wrote,[58] the artists and writers holed up in elegant Swiss villas on the lake, drinking wine, doing drugs, and reading ghost stories aloud, raising their voices to be heard against the beating sound of the rain on the roofs. Is it any wonder that this creative incubation period produced Shelley's classic novel of the era, *Frankenstein*?

That summer was filled with the "carpe diem" spirit. With Europe awash in the revolutionary notions of individual sovereignty—versus monarchical rule—the romantics, inspired by their intellectual hero, French philosopher Jean-Jacques Rousseau, focused on the primacy of their own emotional states. On a quest to encounter the sublime, the artists spent much of the summer expressing feelings of awe and reverence in the face of snowcapped mountain peaks and the translucence of a lake that Mary Shelley described as "blue as the heavens which it reflects."[59] Lord Byron organized hikes up into the Alps with only mules as guides, he and Shelley sailed around the whole of the lake in the midst of torrential winds and rains, and the entire entourage spent their afternoons sitting on Byron's villa overlook, absorbed in the beauty surrounding them.

If their days were spent in religious communion with the natural world, then the nights were taken up with an ecstatic communion with one another. Thanks to copious amounts of alcohol and new forms of liquefied opium, the artists sought out heightened connection, forgoing strictures of stuffy middle-class society in the name of free love—ultimately a much better bargain for the men than for the women, as made evident by Byron's rhetorical response to the announcement of yet another illegitimate child: "Is the brat mine?"[60]

In fact, Lord Byron, arguably the most famous romantic, sets the standard for the archetype we still use today. Referred to as "mad, bad and dangerous to know"[61] by English society because of his famously scandalous affairs with both men and women—not excluding his own stepsister—and revered for his darkly handsome good looks, air of mystery and charm, and irresistible aura of rebellion, Byron exemplified the changing zeitgeist of the era. While the neoclassical hero of the eighteenth century—the Age of Enlightenment—was characterized by his ability to *think*, dazzling society with his lucidity in the burgeoning café culture, and espousing airtight rational and linear arguments, the romantic hero was *emotional*: moody, brooding, and unpredictable.

In response to Descartes, Rousseau famously quipped, "I *felt* before I thought." The romantic hero—or antihero as it were—was a direct response to the oppressiveness of rational thought that had characterized Enlightenment philosophers. The romantic poets—and the Byronic heroes to follow—vacillated between an ecstasy and a despair that shunned all forms of reason; they held themselves apart from society, often harboring some kind of secret or dark past that kept them from being entirely accessible to the rest of the world. A natural outgrowth of the romantic's interest in emotional states—the primacy of the subjective experience over objective truth—was the movement's attraction to the nostalgic and the uncanny.

These and other darkly emotional states—such as suffering and distress—became hallmarks of the romantic, famously portrayed in Goethe's *Sorrows of Young Werther*[62] and in the persona of the "suffering traveler" that Wordsworth, Byron, and other writers of the time embodied. Today we consider suffering to be a deficiency, a "flaw" in our underperforming systems (requiring a fix from pills or shrinks), but the romantics heralded suffering as a necessary experience of life: I suffer therefore I am. The Germans coined the term *Weltschmerz* ("suffering from the world") to describe this sentiment, and the philosopher Isaiah Berlin viewed it as the result of an "unappeasable yearning for unattainable goals."[63]

Such yearning placed the romantic in a state of perpetual opposition to conforming norms. When taken to its extreme, the romantic retreated from all of society and turned to life as a hermit. And if such radical choices proved too difficult to maintain, wealthier romantics simply outsourced the job to the hired help. British aristocrats, for example, started designing their estates with a hermitage on the grounds, keeping a hermit-in-residence on staff. As one account describes, the hermit was expected to live in a "suitable distance" to the main house, to "remain bearded and in a state of picturesque dirtiness . . . peering into the semi-darkness with a little thrill of wonder and excitement."[64] Hermits became emblematic of an authentic life untainted by the corrupting forces of the world. Two centuries later, comedian, actor, and writer Steven Wright put it best: "Hermits have no peer pressure."[65]

Vestiges of these and other elements of romanticism remain with us to this day. One need look no further than our culture's latest craze for vampires—Robert Pattinson, the star of the *Twilight* movie series, has more than a touch of that Byronic curl in his hair—or the iconic figures of rebellion ranging from James Dean to Jim Morrison to Kurt Cobain to Edward Snowden to find traces of the romantic hero in our contemporary culture.

We can place our society's most famous figures into one camp or another. If James Bond is romantic, then the whip-smart deductive reasoning of Sherlock Holmes is not; Humphrey Bogart, clearly Byronic; Tom Cruise, not so much. In business, you can argue that Virgin founder Sir Richard Branson or Apple's Steve Jobs are romantic heroes while leaders such as Bill Gates, Warren Buffett, or GE's Jeffrey Immelt, decidedly, are not. By extension, any figure that stands apart from society, characterized by an aura of mystery and brooding emotion rather than lucidity and rational articulation, belongs in the family of romantic heroes.

In this way, the word "romantic," once primarily descriptive of an artistic movement, has morphed into our popular vernacular. Today we can toss around "romantic" in any number of contexts, and people will nod their heads in a collective understanding. When we think of romantic experiences, most of us imagine something that lifts us out of the doldrums of the daily routine. We imagine the Taj Mahal, tango in Buenos Aires, or a candlelight dinner on the beach. We think of strolling through the Quartier Latin in Paris; "Roman Holidays"; a marriage proposal on the Golden Gate Bridge; or an unexpected delivery of flowers. The tropes of romance have become so familiar that entire genres and subgenres have been developed to express them—everything from the "bodice rippers" of the mass-market paperback world to the more sophisticated work of auteurs such as Woody Allen and *Manhattan*, his love letter to New York, to the famous "rom coms" of the 1930s and 1940s such as *Bringing Up Baby* and *The Philadelphia Story*. The French, masters of romance, have the perfect phrase for it: the *je ne sais quoi*, an inexplicable, intangible, and unquantifiable "something" that transforms our experience of the world.

From the romantic poets of the past to contemporary culture's romantic heroes, the defining characteristics of the romantic have remained more or less the same over the centuries: the elevation of emotion over reason and of the senses over the intellect; an introspective orientation and fascination with the self; hypersensibility and heightened awareness of sentiments and moods; a keen interest in strangers and strangeness; the pose of a contrarian; a belief in community combined with a need for solitude; an appreciation for the sublime, mysterious, and secretive; a sense of awe for nature; a belief in imagination and beauty as pathways to spiritual truths; and the desire to engage the whole self in profound experiences. Connecting all these traits is the quest to live a richer life, one that transcends the boundaries of rationality, social norms, and cognitive and emotional consistency, with everyone and everything full of meaning.[66]

It sounds like a magnificent ride. And yet we struggle. How can we possibly transpose these notions of experiential living to our market society, to our transactional culture of business? Where do we begin?

As with any journey, it is best to start at the very beginning. What elicited our passion for life before we knew about the more "mature" values of money and power? How did our first experiences with work put us in touch with our most romantic selves? Was it the whoosh of the revolving door when you entered your first office? Or the feel of the crepe dress on your legs when you were called in for a promotion? Maybe even the chill of the air-conditioned cubicle felt invigorating after you signed the big client? For many of us, these early thrills have long since gone away. But we can reach them viscerally through our memories, through artful evocation. Close your eyes and cast your mind back: the way the work felt, the way it sounded, and how it made your heart beat faster.

FIRST JOBS, FIRST LOVES

Over the past few years, at parties, in casual conversation, on travels, and at conferences, I always asked the same question: what are your first memories of work? Some people responded briefly, some negatively, but most people considered my question thoughtfully and then started to smile. I smiled, too, as I listened to them. It was impossible not to be touched by such personal confessions. After all, people were sharing stories of their first love.

- *"That cash register sound: duh-ding!"*
- *"The chill of the ice-cream scooper."*
- *"The smell of smoke and grease on my clothes."*
- *"My fingernails were always dirty."*
- *"You could smell the print in the newspaper offices."*
- *"Those little tiny rows in my bookkeeping journals."*
- *"Smeared ink everywhere."*
- *"I always had hay in my hair."*
- *"Tired—so tired you just crawl into bed."*
- *"Popcorn with butter."*
- *"I had such a funny tan from sitting on the dock."*
- *"Pushing that wagon with only three wheels up and down the beach."*
- *"Can I help you? One moment please."*

An investment banker I met at a party in New York started by answering my question simply: "I remember being interested in what my father did." Our conversation was interrupted, and I didn't think I would hear any more from him on the subject, but he sought me out later. "I've been thinking about your question," he said. "It brought up a lot of memories for me." Later that week, we spoke by phone, and he told me this story:

> When I was a child, I couldn't wait to put on a suit and go to an "office," like my dad. Every day, I watched him stand at his bathroom mirror in a white T-shirt, put Vitalis in his hair to slick it down, and then pull one of his ten different suits out of his closet. Most of his suits looked the same to me, but I knew he had each one tailored with "invisible" bespoke details in the cuffs and pockets. Inside, where no

one else could see, there were beautiful silk linings of vibrant colors. He called these his "secret smiles."

After he was dressed, I would walk him to the front door and place his briefcase in his hand. He put on his overcoat and rubber boots over his "dress" shoes. We made the same joke almost every morning of my childhood—a little hand slap of "Gimme Five"—and then I watched him disappear into the dark, cold Connecticut morning.

What was this wondrous land called "the office"? What kind of magical place would warrant such a precise ritual of dress and conduct? I promised myself that one day, I, too, would wake up in the morning and go to an "office." I would enter into the secret society of grown-ups with a uniform of suits in their closets. I would leave on that magical 7:52 train every morning and be transported, along with my dad, to the honored guild of businessmen.

And I have. Today, I wake up every day and get ready to go to work in just the same way. My father passed away five years ago, but I still think of him almost every morning. I still think about all of the different ways I am carrying on something important about his traditions and his values. The office no longer seems magical, but my morning ritual—and the way it evokes the spirit of my father—decidedly does. We still get ready together every single morning. We are, together, men of business.

An older woman I met—a successful head of human resources at a technology firm—told me, suddenly giggling, about the joys she felt in her first years as a babysitter:

I just remember the pure thrill I felt as soon as all three of the neighborhood kids were finally in bed. I would shut the door, wait a few minutes, go back and check, hear the sounds of their breathing, and then, bam, it was like an instant rush. The whole night ahead of me: an entire evening of "adult life." Sitting in my neighbors' living room, stretched out on their couch, my feet up, I just felt everything opening up, everything full of possibility, my job well done, nothing but promise for the night. It was so satisfying.

She stopped herself briefly before continuing.

> Work was so simple then, but the rewards were so profound. How is that possible?

While traveling, I struck up a conversation with a man who owned his own car repair shop. He told me about the first time his father sent him on a dolly "under the chassis":

> I was really little—five or six. It was so dark under there and so quiet all of a sudden. I just felt at peace. It was like I had landed on a different planet, and I was allowed to wander around and see it at my own will. I wanted to know how to navigate it. I wanted to understand what I was seeing. I remember feeling that I desperately wanted to learn.

A police-officer-turned-private-investigator in Texas told me about his father, a marine, and his uncle, a Chicago cop:

> Cops, marines—everywhere I looked, I saw uniforms. These were my models. And those uniforms meant everything. I wasn't sure what I wanted to be. I just knew I wanted to be wearing a uniform.

A miner I met in California reminisced about the first time he went to look for gold with his family. Although he makes his money in the mining industry to this day, he has never tried to sell the gold nuggets he found in his childhood.

> Out there, enjoying it, it was never meant to be a means to an end. It was just an end. And holding gold proves to me that I had all those times. I like to take each nugget out and look at it and remember how I found it. I look at my gold, and I relive each and every experience.

These first memories of work can serve as totems for all of us. We must keep sacred our younger, cherished notions of work and its meaning. We must remember how romantically unfulfilled we felt during all those early experiences, how the promise for more kept us both excited and engaged.

This spirit of *un*-fulfillment is too often missing in our adult work lives. We chase after a job at the "sexy" new start-up or a quick career stint that is the equivalent of a "one-night stand." We put in tremendous effort to acquire these things; we desire them, and then we ultimately possess them because they are attainable. And yet, when all is said and done, such acquisitions often feel hollow. Like the lothario with notches on his bedpost, we begin to quantify that which is most magical and elusive. Sex holds the promise of instant gratification. Romance, however, is like Don Quixote's "impossible dream" or Captain Ahab's obsession with capturing the Great White Whale: the chase for perpetual unfulfillment, with the goalposts constantly moving.

The difference between sex and romance calls to mind French journalist Sophie Fontanel's passionate case for the possibility of a chaste love affair with the world. In her book *The Art of Sleeping Alone*, Fontanel insinuates that the world's most romantic relationships remain platonic.[67] For Plato, the ancient Greek philosopher, love is humans' most powerful tool to contemplate divinity, but that doesn't mean that platonic love must be antierotic. In fact, he distinguishes between "Vulgar Eros" and "Divine Eros," between a self-obsessed, material, and attraction-focused sexuality for the purpose of physical pleasure and reproduction and a more sublime erotic sensibility that transcends the desires of the flesh into the spiritual realm.[68] Similarly, Sigmund Freud defined libido as lust for life, not lust for sex.[69] These conceptions fuel the revelations Fontanel discovered in her sexless love life:

> During the 12 years I didn't have sex, I learned so much. About my body, the role of art in eroticism, the power of dreams, the softness of clothes, the refuge and the importance of elegance . . . But I've learned that most people mainly want to prove that they are sexually functioning, and that's all.

I would argue the same for our life in work. We want to prove that we "function" as well, if not better, than the machines that surround us in our offices. And yet this outlook will only lead us to outsource or full-on automate ourselves when in fact the greatest value we bring to our jobs is our human capacity for imagination.

Time-management studies find that the average American adult devotes just four minutes per day to sex, but more than four hours per day to imaginary worlds, immersing him- or herself in books, movies, video games, and television, or simply daydreaming and fantasizing.[70] As the psychologist Paul Bloom points out in his book *How Pleasure Works*: "Our main leisure activity is, by a long shot, participating in experiences that we know are not real."[71] Our imagination transports us to other lives, to other worlds. This is a quintessentially romantic posture. In fact, Nietzsche, the German philosopher, described the romantic as a person who always wants to be elsewhere.

When I "came out" to the world as a Business Romantic—defending these very pleasures of unfulfillment—I started to meet more and more professionals who "wanted to be elsewhere" while being devoted to the here and now. Some of them were poised to use business to explore their passions while others found great energy and inspiration in committing to their ideals, demanding that business deliver an experience of excellence to both its customers and to its practitioners. Still others were using business to explore deep questions about the nature of our existence together as well as our communal goals.

In the following section, I introduce you to seven Business Romantics working across all aspects and domains of business. The prototypical romantic from the nineteenth century shunned all mainstream culture and norms—standing on a mountaintop with a fiercely guarded solitude—but Business Romantics roll up their sleeves and get down to work, channeling a rebellious spirit and passion into initiatives and provocations that have impact at the workplace, in the boardroom, in our customer experiences, and beyond.

You may recognize yourself in many, if not all, of these portraits. Whether Lovers, Business Travelers, Outsiders, Voices, Guardians, Visioners, or Believers, all of these Business Romantics are seeing and looking for more. All of them are in business for the romance of it all.

CHAPTER 2

Meet the Business Romantics

THE LOVERS

Gastón Frydlewski and Mariquel Waingarten are in love. They love each other, that is clear, but they are also head over heels in love with business. "It all started because I didn't like my shoelaces," Gastón told me. It was 2002, and the twenty-one-year-old business administration graduate found himself constantly complaining about them: the way the ratty knots made his "kicks" look so ugly, the clumsy design of tying and untying, their uniformity amid the craze of personalized sneakers. His frustration with the ubiquitous design rankled him at first, and then, slowly, it started to consume him.

At the time, Gastón, a native of Buenos Aires, was trying to decide his career path. "I met some entrepreneurs, and their creativity triggered a new way of thinking for me. They told me, 'If you have a good idea, the money comes.' I didn't know that was possible. I realized I was passionate about redesigning shoelaces, and I decided to put everything into my goal."

Gastón sketched out preliminary business plans and strategies for a new kind of product. He hired industrial designers to prototype a better shoelace. But, still in his early twenties, he lacked the gravitas necessary to win over the investors.

"They told me I was too young, too inexperienced," Gastón explained. "They thought the idea was crazy, and they weren't comfortable handing money over to me. I needed business experience."

Instead of giving up the dream, he went first to get his master's in finance and then to work for JPMorgan in Buenos Aires.

"I was in mergers and acquisitions for five years—all sorts of crazy deals all over the world, getting real-world business experience. But at the end of every month, I always reinvested the money I made back into my prototypes, my R&D for the shoelace project. I had small goals to keep me committed: in my free time, I would try to do a little bit every week to develop the business."

It was during this period that Gastón met Mariquel. At the time she was a contemporary dancer and photographer as well as a former real-estate developer, busy running her own boutique hotel. With only five suites, Mariquel's hotel was small but enormously popular. Only one year after opening, it was rated on TripAdvisor as the number one boutique hotel in Buenos Aires.

Gastón and Mariquel fell in love and quickly moved in together. Both poised for great success in their respective careers, they might well have stayed the course and enjoyed a life of wealth and privilege like so many of their colleagues. But something was missing. As Mariquel put it, they wanted "to bring something out into the world." They also felt alienated by the corporate culture and what seemed to be—in the digital age—an increasingly byzantine method of R&D and fund-raising. They sensed there was a more exciting aspect to business, but they were certain that it wasn't happening in the hallways of JPMorgan and other conglomerates. They wanted their business to reflect their own life values and choices, not those of a corporation.

"After a year, Gastón and I were growing frustrated," Mariquel recalled. "We were apart for most of the day. Why does work have to mean being away from your family? Why is society structured in this way? We both thought, 'We should change this.' "

At that point, eight years into his prototypes, Gastón's designs for a better shoelace were beginning to take shape. With the help of professional designers, he created a series of rubber-band-like tabs with notches at either end. When "popped" into the holes intended for shoelaces, the design instantly transformed sneakers into "slip-ons." The result of his winning prototype was a sneaker with five zippy little colored bands creating a bridge between the shoelace holes—and no ugly bows.

With enthusiastic feedback on the latest designs, Gastón and Mariquel decided they were ready to quit their jobs and start a business. After getting married, they packed up everything to move to a new city for the official launch of their burgeoning lifestyle brand. Instead of heading straight to Manhattan—ostensibly the "fashion capital of the world"—they opted for the energy of a new business movement happening across the river. "Brooklyn was filled with people with new ideas about how to make business work," Mariquel told me. "These people were building society with new tools, and we wanted to be a part of that."

After walking along the riverfront in the Brooklyn neighborhood of Williamsburg—a series of once-decrepit nineteenth-century factories now rehabbed into chic industrial co-work spaces and studios—the newlyweds knew they had found a home.

In 2011, the new brand launched, infused by inspiration and more than ten years of dedication, but, most powerfully, by romance. The couple named it HICKIES.

Business-cum-revolution is not for the faint of heart. The

two entrepreneurs worked nonstop during that first year, even attempting some moves from the "old-school" business playbook. They met with countless retailers to discuss selling HICKIES in stores, but they were met with either indifference or offers of a completely unsustainable business model.

"The retailers all told us, 'This won't work,' or they said, 'Give me seventy percent of your profit,'" Mariquel recalled, laughing in disbelief. "We said, 'Why is someone from a bricks-and-mortar store—the old world—telling us how things are going to go?'"

Instead, the couple walked away from more traditional forms of retail and investment and turned to a new form of crowd-sourced funding. In 2011, Kickstarter—a software platform enabling artists and entrepreneurs to raise money for their projects—was just becoming popular. The couple spent more than three months creating their campaign.

HICKIES launched on Kickstarter in May 2012. On the first day, they raised $25,000. When they ended the campaign forty-five days later, they had brought in $160,000. All told, four thousand backers—buying ten thousand packs of HICKIES—signed on to support the HICKIES business model. After the success of their launch, the retailer Brookstone invited HICKIES to begin selling through its stores. Between October and December of 2012—in less than three months—they sold two hundred thousand packs. In their first official year of business, HICKIES had already broken even.

Today, only two years later, the company has raised $4.2 million in funding, growing its operations to include a European subsidiary and shipping one million packs of their signature lacing system to fourteen different countries.[1]

"I channel my creativity through business; this is where I express myself," Gastón said. "Because, at the end of the day, business is just another means of 'expression.'"

THE BUSINESS TRAVELERS

In October 2012, artists and entrepreneurs Emily and Adam Harteau packed up a few choice belongings as well as their nearly two-year-old daughter, Colette, and drove their iconic VW Westfalia camper out of Los Angeles. The plan was simple: head south. They started by exploring Baja California and Mexico, then headed farther south to Central and South America. Except for brief stops back in California and in Hawaii to visit family and friends, they committed more than eighteen months to a life of adventure and the great unknown. And they don't see an end to their journey anywhere in sight. Already their daughter calls their camper van "home," and she peacefully falls asleep to the hum of its motor.

Their trip harkens back to the sixties and seventies when such journeys were so emblematic of the countercultural movement in California. Both Emily and Adam grew up traveling and camping—Adam spent large parts of his childhood in a salvaged school bus driving across the western coast of the United States—so this journey is a reconnection to their own childhoods.

What makes their story different, however, is the way the Harteau family is using sympathetic market communities to fund their travel. Emily and Adam document their life on the road through a stunning Instagram feed: Adam is a photographer of note, and his images present a glamorous life of nights spent sleeping under the stars and days spent exploring the wild places that the family encounters. Emily updates their blog with vivid commentary of their travels, and she is currently working on a cookbook based on her VW stove-cooking experiments. The Harteaus' romantic life project has garnered a large audience as desk-bound and travel-hungry followers eagerly await news of the family's latest adventures. In our current age of ubiquitous Starbucks, laptops on the beach, and smartphones in bed, the wanderings of the

Harteau family offer up a subversive and convincing alternative to the good life.

Business has made it all possible. Before their journey, the family started a Kickstarter campaign to raise money for life on the road. When their savings dried up mid-journey, they used their active online community to start a series of Flash sales—dubbed 24-Hour Bazaar—offering their followers an opportunity to engage in their project by buying local handicrafts. (The most recent sale featured hand-woven rugs from the family's travels through rural Peru.)

"Business was not part of the original equation," Emily told me. "But it has become an essential piece of the formula for us to continue life on the road."

Instead of being a means to an end—business to make money—business has become an integral part of the journey. In fact, the Harteaus describe the market exchanges as some of the most powerful community-building aspects of their entire project.

"The 24-Hour Bazaar has allowed us to connect with many people that we would not otherwise have met: artisans have invited us to their homes, we have shared meals, our children play; we have a great relationship with many customers. We are continually humbled by the interest we receive from around the world," Emily said.

It is appropriate then that the Harteaus have named their family project "Our Open Road." Their business model resonates with our nostalgic notions of travel from the past while remaining committed to a more progressive point of view on the tensions between exploration and exploitation—what Emily described to me as "a modern balance of cultural awareness, respect, and understanding for the places that we experience."

Perhaps nothing is more nostalgic than the very vehicle that is carrying them across the world as a family. The VW Westfalia is so revered as a brand that it has generated countless travel blogs, articles, underground marketplaces, YouTube videos, and love letters in the form of fan clubs and societies.

Emily agreed: "The visual weight of our home on wheels is surely the most nostalgic element to the journey. But the entire project is in the spirit of reexamining the American Dream."

Our conventional notion of the American Dream moves from poverty to success to ownership to a role as recognized citizen in the community; the Harteau family's journey, on the other hand, imagines the dream in reverse. By stripping the function of business down to its most essential bones—cultural capital is the only thing that accumulates—the project can never scale. Our Open Road exists entirely in the here and now: there is no five-year plan; there is no strategic growth. Sometimes a family business needs to stay in the family.

"For us, as a family, business and adventure are not mutually exclusive," Emily said. "They are intrinsically intertwined."

THE OUTSIDER

Alexa Clay considers herself a misfit in the world of business. After completing a master's in economic history at Oxford, the thirty-year-old studied models of social entrepreneurship with the nongovernmental organization (NGO) Ashoka, eventually leading to the development of a movement that she and her business partner, Maggie De Pree, dubbed the "League of Intrapreneurs." Alexa immediately felt drawn to the intrapreneurs she met—people attempting to do social innovation work from inside large corporations—and she started to explore the strategies and tactics these types of misfits used to challenge orthodoxies and conformist cultures. Today, all of her wide-ranging projects—she recently completed her first book, *The Misfit Economy*, and she engages in performance pieces such as playing the role of the "Amish Futurist" at tech conferences—are made possible by her outsider status. She is the quintessential provocateur, a hermit who is in the mix, seeing things from a different perspective and revealing possibilities for change.

"As an ethnographic landscape, business has always been really foreign to me," she told me. "It is still strange for me to see the suit and tie, and the more conformist environment. The people I have always identified with are bringing their outsider personalities to their more corporate identities. This is what allows them to innovate."

Innovation driven by outsiders has never been more important, she says. As we make the paradigm shift from the command-and-control centers of twentieth-century models of capitalism—factory work and bureaucracies of white-collar corporations—back into more peer-to-peer marketplaces akin to cottage industries, more opportunities for autonomy, creativity, and romance are opening up. Alexa's work is intended to wake people up to the possibilities for richer and deeper experiences within the culture of business. "We have to reteach ourselves how to be human within our organizations," she insists.

This is not an uncontroversial stance. After all, bringing more humanity to the workplace leaves all of us more vulnerable. We are putting ourselves at risk of a large-scale co-opting of our cultural values. Alexa even references the example of Henry Ford, a "creepy capitalist," as she calls him, who attempted to create a utopian factory community in Brazil that melded capitalist and socialist agendas: "He was very invasive in the private life of his workers. He outlawed vices like smoking and drinking. He encouraged certain lifestyle choices. In Brazil, the people revolted. The settlement that he created failed."

And yet, despite such an example, Alexa argues that this is precisely the moment for taking a risk. Our lives are already so encroached upon by the normative values of capitalism that our only choice is to reveal our fullest selves within this mainstream market culture.

It is easy to imagine how an entrepreneur using Kickstarter or Etsy might bring his or her fullest self to work. After all, their work is as much passion as it is paycheck. But Alexa is trying to catalyze the outsider's romantic agenda in traditional corporate culture as well. Her work with the League of Intrapreneurs is a powerful tool: "The League basically started with twenty conversations with different folks within huge, massive multinational companies. I had an hour-long conversation with each of them that literally left me tingling. They were committed to working in these companies, that was how they chose to make an impact in the world."

Alexa and her partner, Maggie, created salon spaces for these executives to meet and share strategies. These events provide the intrapreneurs with an opportunity to swap stories about the most effective ways to implement change within larger organizations, including conversations about identifying corporate levers of power and finding common cause with the more conservative members of senior management. Most important, the League gives intrapreneurs the permission to build camaraderie with their fellow outsiders.

"There is a real need for intrapreneurs to be connected because they feel so isolated from other people like them. Dave British, manager of social sustainability at Ford Motor Company, for example, is a third-generation autoworker—his grandfather worked on the Ford assembly line—and yet he sees a different future for the company. He's trying to get them to think beyond car manufacturing to embrace a business model of mobility solutions. This requires dismantling the model of 2.2 cars per family and working toward services that are better suited to urban density and more environmentally sustainable initiatives like car sharing," Alexa told me.

Staying true to one's calling—especially in the midst of a larger corporation—will never be an easy task. According to Alexa, people like Dave British are naturally endowed with passion and doggedness, but many of them need management skills to communicate with senior executives: justifying projects with a solid financial picture as well as showing how such endeavors might ultimately scale and integrate back into the organization as a whole.

From the Vodafone executive who identified an external source of funding in the company to create M-Pesa, the now-ubiquitous mobile banking solution in East Africa, to ExxonMobil employees working toward alternative energy solutions, to the Morgan Stanley executives developing microfinance programs: intrapreneurs—outsiders on the inside—are bringing their grand visions for social change to the cubicle.[2]

"If we could all honor our misfit selves, what would the economy look like?" Alexa asked. "How would things be different?"

THE VOICE

As the editorial director of Twitter, Karen Wickre is the official wordsmith—and voice—for the company. This position requires an astute ear for language but also an innate sense of style. Karen knows how to take the temperature of Twitter's culture and respond appropriately.

"I generally have a good sense of what to strive for in each communication in whatever format," she told me. "Some messages are inspirational, some call for a specific action, some need to convey technical information clearly, and some are meant to tease. I know the differences and work to make sure our various communications show them."

Karen arrived in this newly created position of editorial director in October of 2011. The company had around seven hundred employees at the time and "lacked a unified voice," she said. She was tasked with creating cohesiveness for all interactions without squashing Twitter's maverick, bottom-up culture. Some of that work now involves managing a network of more than fifteen blogs, finding the right format for telling a story about the company's news or products. In disaster relief efforts, for example, like the Super Typhoon Haiyan, which struck the Philippines in the fall of 2013, she used Twitter's International Services blog to usher in vital information about free data plans for survivors.

"No two days are quite the same," Karen told me. "My team encompasses several functions including editorial, our own social media outreach, internal communications, company-wide speaking opportunities, and guerrilla projects." In this way, Karen's work as a unifying voice has nothing to do with setting rules or serving as an official gatekeeper to communication channels. "That wouldn't be possible, but it also wouldn't be desirable," she said. If anything, her work is more akin to that of a great restaurant chef orchestrating signature meals in the kitchen.

None of this spontaneous inventiveness would be possible without Karen's training in the liberal arts. "There isn't a day that I don't apply my education on the job," she said. Karen believes that her degrees in the humanities—encompassing literature, art, and history—have shaped her perspective on human nature. Most important, her background in the liberal arts required excellent writing skills, cultivating her appreciation for literary motifs and tonal shifts in mood and perspective. "Humanities grads can draw from a deep, deep well for everything from literary allusion to historical precedents," she points out. "There will always be a need for people who can write, think, and feel."

Karen's personal Twitter feed (@kvox), full of curiosities, artistic sensibility, and whimsy, offers a fitting portrait of the tech worker trained in the liberal arts. She says "Amen" to a reminder not to say "blog" when it should be "blog post," she "binge-comments" on *House of Cards* and other TV blockbusters, and keenly observes other users' Twitter-versatility ("I love how you network even in your tweets.").

As a journeywoman of the tech industry over its last thirty tumultuous years, Karen, 62, is grateful for her perspective on the culture of technology. "One of the greatest benefits of age is that I can marvel. I can enjoy things about the industry that my colleagues can't really appreciate at their age." She expresses a sense of delight and surprise at the way her life has played out. "I never would have predicted that I would be in tech for so long," she said, laughing. "I was a liberal arts major: I didn't have a plan!"

She started her career with the nonprofit Media Alliance, and that led, in the 1990s, to creating print publications and conferences exploring the new technology of the Internet. After writing and editing positions at consumer magazines such as *PC World* and *Computer Life*, and consulting with Sun Microsystems, she heard about an exciting new technology company revolutionizing Internet search. Karen was immediately intrigued, and she jumped on the opportunity when a pre-IPO Google offered her a position in 2002. She described the firm as "astonishing" and "life changing," impressive words from a woman not easily seduced by the siren song of tech culture. Karen was excited by the dramatic scope of Google's ambitions, the firm's sense of adventure, and its drive for excellence.

She spent nine years there serving as a digital "auteur," overseeing more than 150 blogs, and then, later, as the company's official presence on Twitter. Eventually restless for new adventure and surprise—she loves getting to work in the morning just so she can "find out what happened during the night"—Karen made the move to her current position at Twitter. While Google's expansive spirit and the scale of its disruptive abilities inspired her, Twitter makes her appreciate the transcendent beauty of 140-word conversation.

She refers to "only-on-Twitter moments," where people and their disparate cultural spheres unexpectedly collide through the platform. These social collisions are typical of what linguists describe as "phatic" communication: communication for the very sake of communication. Twitter exchanges disseminate information—in the case of disaster relief, for example—but, more often than not, they showcase the art of chitchat. Twitter celebrates the pleasures of witty repartee, provocations, and lightning-speed slices of intimacy. What makes the platform unprecedented is that all this chatter is happening on the global stage. When Bill Gates welcomes Warren Buffett to Twitter by offering up a bridge play, and Bill Clinton quickly chimes in, "What took you so long?" the intimate barbs play out in public. "There's never been another vehicle like it for radical self-expression, real-time information, and mystery," Karen told me. "This is all true because of one central fact: humans power Twitter—and humans are mysterious, earnest, endlessly creative, and creatures of community," she said.

Twitter provides more than 250 million of them with a platform to make their voices heard. So it makes sense that a graduate in the humanities like Karen is at the helm, ensuring that the company's voice feels human, too.

While Twitter is certainly the medium of the moment, Karen makes no attempt to prognosticate. Who knows what the future will hold? "Internet companies don't make a five-year plan!" she quipped. This constant of change—a hallmark of the tech industry—has provided an unlikely home to someone like Karen. As a culture, it is not prone to dwelling in the past—"We don't wax nostalgic about the old days. As long as I work in tech, the story is not written; it's not even ending"—and yet it offers up unexpected moments of poignancy. When Karen reflects on the pace of technological achievements, she is filled with a sense of wonder. She appreciates the human drama of it, the way that technology's relentless innovation pushes us beyond our perceived limitations, rushing us all toward unfathomable futures.

"I was at a thirty-year Mac event, and there was a lot of gray hair. There was a nostalgic feeling—'We were there at the beginning'—but it was not a longing for the past because the other feeling running through the room was, 'It was so primitive back then!' It's unbelievable how far we have come, just unbelievable. All we can do is marvel."

THE GUARDIAN

Ansgar Oberholz is one of these people who seem strangely familiar, someone who instantly makes you feel at home. At forty-two, he is the spiritual guardian of the recent Berlin start-up boom and the veritable heart in the center of its business culture. A serial entrepreneur with extensive experience in the creative industry, Ansgar founded St. Oberholz, a café in Berlin Mitte, the city's hip center, nine years ago. The fact that the café bears his last name indicates how personal this venture is to him. A sign outside the café says NOT EVERY COW CAN BE MILKED, and at the beginning of the tour that he gives me, he insists that the founders in Berlin are not in it for the money. "It's the chance to connect with other like-minded entrepreneurs, to create a culture, *their* culture," he said. In contrast to start-up hubs such as

Silicon Valley, Silicon Alley (New York), Silicon Roundabout (London), and Silicon Beach (Los Angeles), Berlin has not come up with its own catchy brand. It begs to be different. Rather than the latest venture capital round, the scene here boasts its sense of community, its lifestyle flair. Historically, Berlin, unlike the rest of Germany, has been a city where failing is permitted; in fact, it is a badge of honor.

Ansgar told me, "I see Berlin as a lab for politics and the economy, and St. Oberholz is a part of that lab. In the midst of all this change, people are still here, and they are still having ideas."

In Berlin, life is not always easy (LIFE IS NOT A PONY FARM, a sign at the St. Oberholz café reads), which is palpable in Berlin's grim winters, but survival isn't particularly hard either, due to the stubbornly low cost of living, compared to other European capitals. In a country that is much more comfortable with saying no than yes, Berlin's faster metabolism and taste for action is both astonishing and inspiring.

The very retro St. Oberholz café is the mother of this new Berlin vibe. When it appeared on the scene in 2005, it instantly became home to the local start-up community, but also to journalists and artists. Today it is a club that is open for everyone, combining old-world charm with new-economy vim and vigor, grace with pace, and literary impulses with business acumen. This special confluence of cultures is manifest in the rich history of the building itself. The nineteenth-century Aschinger brothers, famous German gastronomy entrepreneurs, ran a two-story restaurant on the site of the St. Oberholz that became a popular venue for Berlin's art and cultural avant-garde scene. Ansgar was attracted by the history and unique "spirit of the place." "You need to respect the history of a location when you build something new," he said.

On the ground and first floors, the St. Oberholz is a café, with spartan tables and a fast and friendly coffee bar, an anomaly in a city where taxi drivers sometimes refuse service if your destination is close by ("You can walk there!"). On the second floor, connected by a magnificent staircase, the building houses several co-location rooms, in which up to forty people can spend their working days. A confusing array of power cords hangs from the ceiling, so that the users can plug in their laptops quickly. Germans would call this *Kabelsalat* (cable salad)—the arrangement doesn't look efficient, but it is; and in addition, it gives the space a certain grittiness that is, well, very Berlin. The walls feature a Berlin artist's chalk drawings that mimic the simplistic character of info graphics and fake transience, with the chalk covered by a spray that protects it from being erased. It serves as a metaphor for Berlin and for places like the St. Oberholz: this playful reoccupation of history through artistic means, the coexistence of historical burden and youthful irreverence. Like its founding father, the St. Oberholz is not trying to be cool. It is an earnest place full of warm irony.

"I try to remember why we started all this," Ansgar said. "That first magic moment when I said, 'I have to do that! I have to take that risk!' I am always trying to get back to that place even now, nine years later. I delegate as much as I can, and that helps me keep things romantic. If you do everything yourself, you will definitely kill the romance."

Ansgar is deeply entrenched in the world of incubators and valuations, and yet, like the Great Gatsby, he conceives of an "incorruptible dream." He's not a naive daydreamer, a business skeptic, or an idealist. He's a Business Romantic.

"The heart is very important," he told me. "There are not a lot of questions you can answer with a calculator. My heart says, 'We should find a different location,' and then we need to make a plan and check to see if the numbers work. There is a lot of tension between the calculator and the heart, but you need both in a strategic plan."

And when you watch him roam through his café and chat with his customers, you do indeed get the impression that his heart is on display. All of the ideas born at St. Oberholz are his ideas, but more as pieces of an imaginary exhibition of which he is the temporary curator than as "assets" that he owns. On the St. Oberholz website, Ansgar runs a blog with "lost and found" posts, each one beginning with the line "Yesterday, somebody forgot . . ." He proceeds to describe forgotten items in great detail like a cabinet of curiosities: a napkin with a phone number for "the cute bartender in the black shirt," a ring left behind after a couple's dramatic fight, or tickets for seeing an appearance of the pope in Berlin's Olympic stadium.

Ansgar also founded a publishing company that publishes niche books for "nonreaders," which can be read "during meetings, on trains, buses, planes, on the john, in bed, in line, during routine sex, on the dance floor, instead of small talk." A couple of years ago, Ansgar captured the history, the story, of St. Oberholz in a nonfiction narrative book that was received with wide praise in his native Germany, "but didn't sell well," as he admits with a smile. Not every cow can be milked.

THE VISIONER

Priya Parker, age thirty-one, remembers exactly where she was when she broke down in a quarter-life health crisis. "After completing a very competitive summer internship, I fainted on a plane," she told me. Priya, at the time enrolled in a dual-degree program at the Harvard Kennedy School and MIT's Sloan School of Management, was suffering from extreme exhaustion, getting caught up in life's "shoulds" and losing touch with the ideas and projects that energized her. Her body, and her spirit, simply couldn't take it anymore. When she went to see a doctor, he told her, "You've been on 'war footing' throughout your twenties, and your army simply ran out of supplies."

"It felt as if almost everyone in my dual-degree program wanted to work for either McKinsey or Goldman Sachs when they graduated," Priya recalled. "But they didn't start the program with those goals.

"I began to feel that I should want those things, too, but they didn't really resonate with me. There was an enormous amount of subtle pressure to interview for those jobs, as if they were the only way to be successful." She continued, "The culture and habits of many of these institutions went against my grain. At MIT, all the buildings were known by their numbers (E-52, E-53, etc.) and not by their names. I could never seem to remember the numbers for any of the buildings, even after a year. To name a building after a number rather than a story or a person went against the way my mind remembered and everything I valued."

Priya's doctor suggested she take some time off from her studies and focus on healing her body and reconnecting with pursuits that brought her pleasure. For the next four months, she did not let herself make any plans. Instead, she practiced living moment to moment—a decidedly difficult thing to do for a high-performing student and worker. When she felt a genuine desire to work or read something, she pursued it. When she felt that familiar sense of anxious obligation—the "I should do this" feeling—she trained herself to stop listening to it. As a result of her mindfulness, she was able to get her health back over the four months. The idea behind this period was to connect with what she really wanted and believed in, and to apply her talents and skills to whatever that might be. The days quickly shaped themselves around her renewed commitment to her dancing, more time spent with friends, and, most important, a unique skill set she had been pursuing in between her studies: visioning.

Priya told me: "In my first seven years after college, I was deeply immersed in conflict resolution, race relations, and diplomacy both here in the U.S. as well as in India and the Middle East. That led naturally to my interest in helping people articulate a vision for their future."

She started her work with roommates and friends, creating an experience that allowed them to articulate goals and passions. During her four-month sabbatical from school, she realized that this work was actually the source of her greatest excitement and sense of purpose. She decided she would spend her remaining time in her dual-degree program studying to become a master at her own practice of visioning.

"I was able to redesign my program into a focused set of classes to develop my skills. I used all of my previous training in Sustained Dialogue and conflict resolution, and I started to take courses in everything from 'presencing' to intergenerational relationships to budgeting to transformations in society and politics."

But achieving mastery is never merely a matter of study. Priya knew the most important piece of her burgeoning practice was practice. For that reason, she also spent the remainder of her graduate school tenure running as many "Visioning Labs" as possible. Over the course of the year, she ran a lab in her apartment almost every day from three to seven P.M.

"Give me three or four hours in a room, and I know exactly what I want to do," Priya said. "I have a specific skill." During that time, she began to be approached by local organizations to apply that skill to teams and companies.

She ended up writing her graduate thesis on her classmates: members of Gen Y pursuing two master's degrees—business and policy—as a proxy for high achievers of the next generation. Her pool of interviewees tended to be biracial or multicultural, internationally minded, and high achieving. "They have a sense of direction," Priya told me, "but they are afraid to commit to anything." She described their driving fear as FOMO, or Fear-Of-Missing-Out, because they are always waiting for the best option. As one subject told her, "Choosing one door to walk through means all the other doors close, and there's no ability to return back to that path. So rather than actually go through any doorway, it's better to stand in the atrium and gaze." The goal for this cohort, to borrow from business-speak, was to "optimize" for everything and commit to nothing.

Today, Priya's visioning career has grown and developed through her company, Thrive Labs, and she still sees variations of this same type—optimizers and recovering optimizers—in almost every single one of her clients.

"Once people identify what they really care about, they can have breakthroughs," she told me. "But so many people I work with have forgotten or been disconnected from the moments in life that made them feel fully alive."

To date, she has conducted hundreds of Visioning Labs, with organizations and leaders all over the world. In addition to conversations, Priya also brings exercises to her labs such as writing, mapping theories of change, and embodied cognition: helping her clients experience the connection between their physical and mental states.

Every lab that she conducts is driven by a single question: what is the biggest need in the world that you might have the passion and the capacity to address? As she explained it, this inquiry is intended to move her clients into action. The culmination of each lab is a carefully crafted personal manifesto: a vision statement for purpose-driven individuals and organizations.[3]

"We spend the majority of our waking lives at work," she told me. "We should be building leaders and organizations that are deeply clear on why they're doing what they're doing and embedding it into the core of their work." In order to do this effectively, Priya works to identify their obstacles. Once a person or company has articulated a sense of purpose, the lab's exercises help reveal impediments to it. The two biggest offenders are often mindless use of technology and prioritizing productivity over purpose.

We need to call these habits into question, she told me, especially when we feel drained or uninspired. After all, some experiences are most meaningful when they are completely unoptimized, when they are inefficient, and maybe even unproductive.

THE BELIEVER

Scott Friesen, forty, is the director of consumer insights and marketing analytics at Ulta Beauty, a cosmetics and beauty retailer based in the Chicago area. His area of expertise, data analytics, could hardly be considered to be romantic. But when I described the character profile to him, he said, "That's me!" He throws himself into work, embracing new hires, new customers, new projects, and new ideas. He refuses to succumb to tit-for-tat office politics and other transactional mechanisms. For Scott, his work is an attempt to connect his employees and customers to an experience of excellence.

In 2004, he began his career out of Columbia Business School with a position at the big-box retailer Best Buy in Minneapolis. Although he never initially aspired to be a corporate executive, he fell in love with his new company. "When I first got to Best Buy, it was a place worth fighting for," Scott told me. "It was an egalitarian culture: not defined by education or past experience but by an individual's contribution. I believe in capitalism, of course, but it has to be done 'in service of' something bigger.

Within the system dynamic at Best Buy, that 'in service of' was real."

Slowly, over a period of four or five years, however, the mood at Best Buy began to change, becoming "more transactional," Scott recalled. The company was struggling to shift from a bricks-and-mortar retail outlet to a more expansive online distributor. Although Best Buy had a history of strong customer service, the senior management started putting myopic focus on the short term. The goal was now less about adding real consumer value and more about making the quarterly number. This change was reflected in the office culture. When Scott first arrived, he was fond of Best Buy's celebrated tradition of giving stock options to employees with the most innovative ideas, regardless of position.

"Suddenly they just stopped the stock option giveaway without announcing it or saying anything about it," he said. "Later, it was revealed that a few executives were awarded compensation packages of millions of dollars."

According to Scott, the shifts in mood at the company were dramatic. "It went from 'We're all in this together' to 'We're not in this at all. You are in this alone . . .' The message was clear: some people are going to be paid millions of dollars, and a lot of the rest of you are going to lose your jobs."

Scott told me this was the moment he started to re-evaluate his commitments, his greater narrative. I asked him if he felt cynical about business and, specifically, corporate America, considering this was his first experience after graduate school.

"Oh no!" he replied. "Even today, I still absolutely believe in the experiences that Best Buy can bring to its customers. When a student has a term paper due, and it's eight hours until class and suddenly his hard drive crashes, it's a very real thing that he can call Best Buy's Geek Squad—the tech support—and they can get his paper off the hard drive. Whether we allow a soldier to speak to his family on Skype or a family to enjoy watching a movie together, the experiences Best Buy brings are real when they are done right."

Scott sees only potential, holding his industry and himself up to the highest standards. He believes that we all deserve business that is "done right." In the midst of company changes, his love for Best Buy as an institution, however, faded. He turned his ideals of excellence, honor, and loyalty toward his immediate teammates and colleagues.

"I often heard that when men went to war, they went for ideals of national patriotism, but they stayed for the person to the right or the left. I felt the same way, in a less dramatic sense, of my own job. I was no longer fighting for Best Buy; I was fighting for the thirty people on my team, and my boss, and all the people who worked underneath us."

Scott ultimately decided to leave Best Buy for Ulta. But, despite the departure from this first home and foundational experience, his personal missions and goals have only grown richer and more satisfying.

"I love the experience of watching people on my team absorb something new and then pass it on to someone else. Ultimately, I want to take everything I'm learning here and bring it to the classroom. It's the experience of teaching and learning that makes business so rewarding for me."

And then there is just the simple joy of a meaningful creation. Like all romantics, Scott takes great pleasure in his work when it reveals an innate beauty and order.

"At Ulta, a really great day for me sometimes simply involves taking out a data visualization I am proud of and looking at it. It might only have intrinsic value, but it's a piece of work I just love. Often, I don't even need anyone else to see it. I just know it's valuable."

Lovers, Business Travelers, or Outsiders; Voices, Guardians, Visioners, or Believers: Business Romantics are out there, and their numbers are growing. My conversations with these and other romantics in the field led me to wonder about our collective notions of work. Do we, as Business Romantics, have a platonic conception of work? Do we have a romantic "template"? I decided to test this idea in the most baldly transactional of all online marketplaces: the classified ads of Craigslist.

SEEKING BUSINESS ROMANTIC TO JOIN OUR TEAM

Reporting to the CEO, the Business Romantic will help colleagues, customers, partners, and society at large see the beauty of the business world with fresh eyes. Embracing hope as a strategy, the Business Romantic presents cohesive narratives that make sense of ever more complex and fragmented workplace and market conversations. Instead of focusing on assets and return-on-investment, the Business Romantic exposes the hidden treasures of business and delivers return-on-community. The Business Romantic develops, designs, and implements "acts of significance" that restore nostalgic trust in business as the most impactful human enterprise and provide internal and external audiences with brand and workplace experiences rich with meaning, delight, and fun. We're looking for a self-starter with strong entrepreneurial drive, exquisite taste, and a proven track record of managing the immeasurable.

Specific responsibilities will include but are not limited to:

- ✓ Carving out spaces for the artful and playful at work
- ✓ Elevating day-to-day interactions and transactions to experiences "greater than ourselves"
- ✓ Finding meaning in the seemingly mundane
- ✓ Shifting organizational and brand cultures from utilitarian/transactional to generous/transcendent
- ✓ Creating zones of discomfort and "critical events" that replace convenience with friction
- ✓ Doing things for no reason and taking part in joyous, aimless activities such as mystery meetings or result-free conversations
- ✓ Conceiving of and implementing passion projects
- ✓ Going on regular hikes with the CFO
- ✓ Providing leadership to other romantics at the workplace

Qualifications:

- ✓ 5–10 years of experience in romanticizing business, ideally in various industries
- ✓ Ability to instantly fall in love with ideas, memes, and projects
- ✓ Ability to balance Big Data with Big Intuition
- ✓ Comfortable with high levels of ambiguity and unpredictability
- ✓ Comfortable with giving chance a chance
- ✓ Excels at empathy and generosity
- ✓ Vivid imagination and appreciation for beauty
- ✓ Ability to enable and cultivate deep connections

- ✓ Risk taker and adventurer
- ✓ Obsessive attention to details (because fixing a typo can change the world)
- ✓ Ability to understand routine activities as acts of love
- ✓ Ability to have secrets and keep them
- ✓ Exceptional devotion to the unknown
- ✓ Favors authenticity over truth
- ✓ Appreciation for romantic comedies
- ✓ Extensive knowledge of English and German romantic literature, experience in writing poems a plus
- ✓ If you match this profile or are romantic in ways we could not imagine, we look forward to hearing from you.

In less than a week, I received more than one hundred job applications:

- *"I'm going to guess that sending a résumé to this is not exactly what you are looking for. So what are you looking for? Outside of your description, of course."*
- *"Dear Business Romantic, I am fascinated and intrigued by this job description. It articulates the possibility of expressing barely limned and subtle ideas of reality. What would that look like in the daily workplace?"*
- *"The idea of working as a Business Romantic in a consultancy is extremely exciting. Thank you for posting this."*
- *"I am quite intrigued as I don't really understand 100 percent of what you are looking for, but from what I can see, you are interested in the more 'metaphysical' aspect of business/work. More about how people interact, how to make it a positive work environment, how to find deeper meaning in the interaction with coworkers, clients, and*

so on and so forth. You are welcome to correct me if I misunderstood."

- *"Why can't work be like this more often?"*
- *"This ad makes me sad. I'll never find something like this in real life."*

It's true: Business Romantics will not likely find a job like this today, but who's to say about the future? If we can establish a foothold, perhaps one day such a job, such an organization, will exist, and romantics will thrive, not just for a moment, but over the course of an entire career. Until then, what can we do to keep the romance alive? And how do we protect it against our enemies?

KEEPING THE FLAME ALIGHT

The greatest threat to Business Romantics is cynicism. Cynicism is the *déformation professionelle* of businesspeople. Whether it is buying out companies with massive debt that the acquired firm has to carry, "shorting" expected losses in the stock markets, selling home loans to people who cannot afford them, installing a Corporate Social Responsibility program to eclipse ethical shortcomings, or misleading an employee to believe he's up for promotion just to exploit his motivation and commitment—the cynical impulse often finds a welcoming home in business. As Oscar Wilde said, "A cynic knows the price of everything and the value of nothing." The modus operandi of business, its natural decision-making impetus, is to narrow the playing field until only the most explicit option is left, and everything implicit (mysterious, dangerous, and risky) is eliminated. Business formalizes, automates, standardizes, and serializes our decisions and experiences until they represent the thinnest stratum of humanity. Business craves certainty and seeks to constantly mitigate risk. When the sublime is too ephemeral, and a curious, open mind too vulnerable, cynicism offers a safe harbor—in the boardroom, in meetings, in the employee cafeteria, and at the work desk. To quote the investor Gordon Gekko from the movie *Wall Street*, "If you need a friend, get a dog."

Cynicism is often the disguise of the frustrated idealist. When you scale cynicism, however, you squeeze every ounce of bitterness out of it, and it becomes a cold engine of indifference where "making a difference" no longer matters. In business and beyond, the cynic is the fundamental nonbeliever. I have often heard the common business saying "Hope is not a strategy." That always struck me as the pinnacle of cynicism. Without hope, you have nothing.

Philosopher Peter Sloterdijk describes the cynic as an "enlightened idealist":[4] he is painfully aware of the gulf between societies', organizations', or individuals' ideals and their practices, and assumes it as the natural state in which to operate. From the Greek father of cynic philosophy, Diogenes, to Niccolò Machiavelli and Richard Nixon, to the mantra "there's no such thing as a free lunch," cynicism's considerable track record is uncontested. Some might argue that our current age of discontent renders cynicism the only remaining logical worldview.

The cynic is the antagonist to the romantic protagonist: the cynic views the CEO's memo to all employees as an obvious attempt to sugarcoat organizational challenges by using boilerplate motivational clichés; the romantic appreciates the gesture and is interested in the meaning between the lines. The cynic dreads the Monday-morning staff meeting as a wasteful forum for fostering a false sense of community; the romantic values it as an opportunity to enact and strengthen the community's narrative. The contrast between the two becomes even more apparent in travel and cultural experiences: the cynic views Los Angeles as a hyper-commercial suburban shopping mall erected between freeways; the romantic appreciates it as a "blue screen" that he or she can program with his or her own stories according to ever-changing needs. The cynic mocks blockbuster movies, books, and bands as pretentious simulacra of manufactured emotions; the romantic believes they are totems of the universal human spirit. The cynic mistrusts any experience that is supposed to get under his skin. The romantic, at the very least, has goose bumps.

As romantics, we recognize that our vision—our flame—has the power to light entire departments, companies, even a greater movement, but we also know that it is fragile. It must be cared for and protected, as bureaucracy and boredom can slowly extinguish it. One day, we wake up and find that the flame is out. When this happens, when it feels like the "romance has gone"—from our relationships, our travels, our sports, our brands, and our work—what we are actually lamenting is the commodification, the economization of something we considered rare, precious, and (to some degree) inexplicable. When romance is gone, transaction begins. As Rafael Ramírez, professor at the Oxford Business School and a leading scholar on the aesthetics ("attractiveness or repulsion") of business,[5] put it, "What is the ROI [return-on-investment] on aesthetics? When you quantify love, you get prostitution."

David Kim, professor of religious studies at Connecticut College, has done significant research on our changing needs within the workplace and society at large. "The romantic, in the traditional sense, believes that life should be all-consuming in the most satisfying and robust way," Kim told me. "There is an emphasis on the soulful aspects of experience." Of course, Kim reflected, such soulfulness is often accompanied by anguish, strain, and heartache. It is these emotions, much more than optimism or a sense of personal happiness, that infuse a romantic's life with a sense of meaning. "They help clarify the romantic's quest, the constant searching or chase to find the heightened conscious experience," Kim said.

If cynicism is our greatest external threat, Kim's insights call to mind our most pressing threat from within ourselves. As romantics, we seek out spaces for the heightened experience, but as men and women of business we are also laying the groundwork for a career. It's one thing to have a series of transcendent moments—like so many pearls on a string—but we want to create a life for ourselves that has forward momentum and a narrative whole. How exactly do we navigate between these two impulses? What do we do to light the flame of romance, and how do we ensure that this same flame stays alight throughout our entire career?

The following Rules of Enchantment provide the framework for addressing these questions. Whether you need to find ways of sanctifying the company you keep, rituals to provide comfort and security in the midst of uncharted waters, or tricks for disarming the cynics and keeping the magic of discovery, mystery, and possibility alive, all Business Romantics can benefit from these rules. They are not based on typical business case studies, and they don't attempt to problem-solve; they won't provide silver bullets, and they do nothing (or not much) for your productivity. Instead, they will challenge you to seek out new perspectives, to value your own idiosyncratic intuitions and emotions, to embrace conflict and friction, and to celebrate your humanity. They will help you lead a more wonderful life in and with business.

As Business Romantics, it's time to reclaim our space. The Rules of Enchantment define the latitude and longitude of our territory. They communicate what we stand for and how we want to live in the world—at work, at home, in our communities, and on the map of our souls.

PART TWO

Rules of Enchantment

CHAPTER 3

Find the Big in the Small

I waited for something and something died / so
I waited for nothing and nothing arrived.

—VILLAGERS, *Nothing Arrived*

On June 1, 2013, an unusual social experiment took place all around New York City. Volunteers handing out name tags reading HELLO! MY NAME IS ______ flooded the city's streets. By midday, hundreds of thousands of New Yorkers were wearing the name tags on their shirts. Complete strangers were transformed into intimates: "Bob," "Khalil," "Rocío," or "Camille." This experiment—dubbed Name Tag Day—was designed to bring surprise and delight to everyone in the city. Sometimes happiness is as simple as hearing your own name.

Such examples of small gestures do not pretend to advance our productivity or efficiency, but they ground us in our space and place, reminding us of our shared experience and our humanity. They make our lives better in small, thoughtful ways.

Going through immigration at Shanghai's Pudong International Airport offers another good example. Travelers are asked to rate the immigration officer's service by pressing one of four buttons. The button for "extremely satisfied" shows a big smiley face visible to both you and the officer who assisted you. This presents a small but significant moment of unexpected and sometimes awkward humor. It is discomfiting to assess a complete stranger's performance right in front of him or her—even more awkward to have that assessment displayed for both of you to see. The fact that the assessment is communicated with a cartoon smiley face contributes to this surreal moment, especially when the smiley face is juxtaposed with the earnest expressions on the face of the immigration officer. Shanghai airport authorities created this experiment after receiving reports of poor airport experiences from frustrated travelers. The interaction was probably primarily designed to gather data, but its greater effect is to carve out a moment of subversive delight in the tedium of the modern traveler's day.

Sometimes the best design is the absence of design. Instead of a high-tech touch screen or a similarly ambitious device meant to ease interactions, the low-tech approach often garners the most positive reactions. Hong Kong Airport, for instance, simply presents the traveler with a box of mints—"with compliments from the Hong Kong International Airport Authority." The mint-powered moment suggests that enhancing the customer experience can be quite simple and inexpensive, and it goes a long way to humanize the all-too-inhumane experience of travel. A child could have come up with it, and this is exactly why it works. The mints create a "delight surplus"—delight in excess of functionality.

These modest moments of intimacy also have a vital role to play in our hospitals. Janet Dugan, a healthcare architect, took inspiration from her recent experience having an MRI (Magnetic Resonance Image) scan. While she was lying still and waiting, she noticed a small mirror that had been placed below the head support piece. It was angled so that she could see through the barrel to the radiology technician and make eye contact with him.

"What a small thing," she told me. "And yet what a difference it made. I felt less alone. I was connected to another person at the very moment I needed support. And even though I'm not claustrophobic, it calmed me some to be able to see out of the barrel—to have a vista in front of me that was deeper than ten inches. I couldn't tell you what color the floor was or if the ceiling was acoustical tile or if the exterior skin was precast or curtain wall. I didn't register any of those things in my journey as a patient. What I did register was that the technician was friendly and that the nurse went out of her way to make me laugh. Don't get me wrong," Dugan concluded, "I firmly believe in the power of design to contribute to the healing process—that architecture can shape events and transform lives. But that day, in that experience, the thing that really gave me comfort was a tiny mirror about as big as a Band-Aid."

We can carve out our own simple but delightful moments with others in our offices and work lives. Helen Dimoff, director of communications at NBBJ, an architecture and design firm, came up with the idea for a voice-mail viral story. This was in 2000 before voice mail was integrated with e-mail. Back then, voice-mail systems allowed users to add a new message to the beginning of received messages before forwarding them on to someone else. Dimoff recognized that the entire system was ripe for a storytelling opportunity. She started with the ending, recording "And they lived happily ever after," before forwarding the voice mail on to a colleague. That colleague then added something like "Just as the boat exploded, they dived into the icy waters," before forwarding it on to someone else in the office.

"The voice mail ended up going to dozens and dozens of colleagues and featured all sorts of heroes and villains in a James Bond–esque madcap adventure," Dimoff recalled. "It wound its way through the firm over about a month and would come back to people's voice mail more than once. We finally transcribed the whole story so we could all read the entire saga."

The U.S. headquarters of AstraZeneca, a biopharmaceutical company in Wilmington, Delaware, adds a small touch of whimsy to its lobby. When visitors walk in the door, they are greeted by unusual signage at the reception desk that reads "DIRECTOR OF FIRST IMPRESSIONS." The sign surprises guests by calling attention to the absurdity inherent in all corporate protocol. It prompts an intimate and authentic bond—a wink-wink—between the company and the visitor. Chris Young, the founder of the Rainmaker Group, is said to have invented the title, and a growing number of companies are now using it instead of "receptionist."[1]

Business leaders can create these delightfully strange moments, too. Guy Laliberté, CEO of Cirque du Soleil, did just that when he hired a personal clown to accompany him to company meetings and mirror his presentations, standing to his left. (Whether you find clowns funny or creepy, the mere presence of this gesticulating performer would certainly be memorable.)

What do all these things have in common? They use a light touch, and often a wry sense of humor, to remind us of the beauty of the world. They are humble in that they don't attempt to take on the big problems. Rather, they exist to gently disrupt. They remind us to look up, take a breath, and direct our attention back out toward the world. Delight is their only ROI.

Business Romantics can use this design imperative when approaching all aspects of office life, especially socializing. I've always wondered why certain people walk into the office and say hello to everybody whereas others refuse to even respond. A third group is kind and polite but "friendly-on-demand." They want to engage but are constrained by what they believe to be the economics of attention: the commonly held belief that too much social interaction can seriously hamper productivity. There just isn't enough time in the day, they seem to think, to say hi to everybody, to smile at everybody who has earned a smile, to engage in a brief conversation. Attention is a zero-sum game, apparently, so they'd better hoard their limited supply.

Business Romantics recognize this is a mistake. A 2013 study by the London School of Economics claims that paid work continues to be "negatively correlated with happiness," in keeping with previous research that had shown similar results.[2] Comparing it with other individual activities such as sports, entertainment, or travel, the new study found that paid work was ranked lower than any of the other thirty-nine specific activities the survey respondents quantified, with the exception of being sick in bed. According to the researchers, there is only one aspect of work that results in happiness levels similar to those experienced away from work: casual interactions with colleagues—in other words, socializing at work.

So if the best way to be happy at work is to chat with your colleagues, why aren't we encouraging more socializing? Well, because it's *business*. And business, for the most part, still operates under the principles of efficiency or productivity gains. To be fair, when it comes to customer experiences, especially at the point-of-sale, companies have long understood the power of casual micro-interactions: some small talk here, a fleeting smile there. But at the workplace and in our employee experience? Not so much.

Business leaders with a penchant for romance are starting to see things differently. Some firms are radically rethinking the design of their workplace, calling into question even the most basic staples of office life such as the desk and the chair.

The flagship feature in the New York office of advertising agency Barbarian Group is a 23,000-square-foot "endless table" that runs continuously through and around the twenty-three-thousand-square-foot space. Also called "Superdesk," the table gives all of the 125 employees—including chairman Benjamin Palmer—a place to work, as well as providing meeting rooms and a communal space under seven intricate archways that punctuate the Superdesk's flow. Architect Clive Wilkinson said he wanted to create an environment that would enhance cohesion and bring the feel of an urban plaza to the workplace. The result is practical (more flexibility and collaboration for a largely paperless workforce), but also a tangible expression of the firm's creative gusto. The table—quite literally—provides endless space for the company's stories to unfold.

Many businesses pursue a workplace strategy of either the carrot or the stick: Google has mastered the carrot and created attractive work campuses and perks that are an antidote to more uniform corporate cultures. Yahoo! or Hewlett-Packard didn't want to rely on attraction alone and implemented a nontelecommute policy, requiring most employees to put in a certain amount of face time at the office.[3] Open-source code-sharing and developer community GitHub goes beyond these two options by arguing that the office no longer has a role to play in conventional productivity. For Scott Chacon, cofounder and CIO of GitHub, the "headquarters" is primarily a social arena, not a productivity hub.[4] Almost all of the staff works remotely. In its early years, the company didn't even have a physical office. GitHub also has rules such as "Hours are bullshit," "People choose their own work," and "No meetings" that question common workplace orthodoxies.[5]

GitHub's definition of work space is extensive. Chacon points to an experiment called GitHub Destinations, where the company rents an Airbnb apartment in a location to which people have always wanted to go, say, Tuscany or Montevideo. "They choose to live there for a month and have serendipitous interactions because you can get just as much done there as you could anywhere else," he told me.[6]

Ara Katz, a Los Angeles–based film producer and entrepreneur, and the head of creative at subscription commerce start-up BeachMint, concurs that a good work experience is less about bland company values and manifestos and more about small moments of intimacy, humor, and pleasure. She stresses the importance of play: "I think of companies as school playgrounds. All companies need to provide some playing space. That can be in the form of an incubator or an innovation channel. There has to be space to explore and to grow. If you don't have someplace to play, then there is bound to be a backlash."

Opportunities for such moments of play exist in the most mundane places; perhaps the most abundant of all is our day-to-day e-mail communication. A 2012 study by the McKinsey Global Institute states that skilled office workers spend more than a quarter of each working day writing and responding to e-mails.[7] But if we want to counter the tyranny of productivity with a human touch, then we must refrain from signing e-mails with only our initials, not to mention using abbreviations such as "Thx" or "LOL" (which UK prime minister David Cameron famously mistook for "Lots of Love"[8]). "I am busy," is what we want to convey, "in fact, so busy that I don't have time to write my full name" (even if it's as short as Tim or Ava). And increasingly, we no longer start our e-mails by addressing the recipient by name. We just abruptly launch into the content. (A colleague stopped addressing me by name in e-mails on her second day at work!) Again, all this seems to reinforce the impression that we are hyperefficient, hyperrational, just-the-facts, it's-all-business productivity machines. Time equals money—so much so that one salesperson once told me about prospective customers: "The longer the e-mail, the less budget they have." That might be true, but with this self-imposed protocol of brevity, we foment a culture of artificial busyness and contribute to the demise of unexpected but essential relationship moments. At least we're not abbreviating the names of those we write to yet (if you are aware of examples that suggest otherwise, please share them with me in an e-mail that opens with "Dear Tim"), but wouldn't it be more romantic to spell out our own names and not just close e-mails with initials or one letter?

Benjamin Ward, a music producer and friend of mine, started an experiment: he began each business e-mail with "Dear" and ended it with his full name (including his middle name, Anthony) and a flowery closing line ("warmest regards and best wishes"). Benjamin even made an effort to ask about the other person's family and covered a few other light items in his opening paragraph before shifting to the business topics to be addressed. Soon after his experiment started, his responders reciprocated with a warmer and more personal tone. At some point, he began to receive unprompted e-mails from contacts that seemed to be emulating his friendlier style. When one e-mail arrived, addressed to "Dear Benjamin," from a business partner who had previously been notoriously terse in e-mails, he knew human touch had prevailed over the regime of "lean."

By the standards of efficiency, Benjamin's practice may count as excessive. By the standards of empathy, however, this kind of thoughtful acknowledgment creates lasting ties. The relationship expert John M. Gottman—renowned for his more than forty years of research on marriages—contends that good relationships are not about clear communication but small moments of attachment and intimacy.[9]

The performance artist Miranda July conducted an entire piece called *We Think Alone* that used e-mail as a magical window to the world of others. She asked for private and ordinary e-mail contributions from artists and celebrities such as actresses Lena Dunham and Kirsten Dunst. I subscribed and enjoyed the subversive thrill of opening a personal e-mail written by a famous stranger, and enjoyed it even more knowing that I was in the company of tens of thousands of other subscribers on the list.

In a statement on her website, July explained, "I'm always trying to get my friends to forward me e-mails they've sent to other people—to their mom, their boyfriend, their agent—the more mundane, the better. How they comport themselves in e-mail is so intimate, almost obscene—a glimpse of them from their own point of view."[10] July's project gently probes at the tensions between our private and (carefully cultivated) public personas. It was touching to read the most boring e-mails precisely because they were written to someone else: "I am walking the dog. Xo." It is the sense of peering into someone's life, especially someone who is so "known," so public. It was a little slice of extravagantly mediated intimacy.

Rowan Gormley, the founder and CEO of the online wine service Naked Wines, decided to play with this concept in his business e-mails. Weary of yet another e-mail blast to potential customers, he decided to try something different when he launched Naked Wines: remove all mention of "new," "free," "unique," "value," "special offer," and any other marketing jargon. Instead, he would write to the customers as if writing to a friend in a casual, witty, and friendly way, without any "call for action." Then he tested this version against a more traditional marketing e-mail to see which one was more effective. The response rate of the personal e-mail was much higher than that of the traditional one. "Isn't it amazing how simply being yourself is the best value proposition?" Gormley told me. "And isn't that comforting?"

I used this design principle of "being yourself" in a dinner I cohosted with Priya Parker, one of the Business Romantics from chapter two. Priya and I invited fifteen conference attendees at a World Economic Forum summit in Abu Dhabi to a dinner we called "15 Toasts." Our guests were leaders from business, government, and civil society, most of them strangers to each other. Priya's husband, the writer Anand Giridharadas, came up with the format for the evening: each guest had to offer a toast at some point, addressing the question "What is a good life?" As a special twist—providing a strong incentive to keep the round of toasts moving—the last person had to sing their toast to the room. After we welcomed our guests, the toasts started slowly and rather formally, but then quickly changed in tone. The third speaker burst into tears as she shared what a good life meant to her, and another guest wept while he talked about the recent passing of a friend. The last one to raise his glass was the owner of a well-known brand and marketing firm, and he concluded the evening with the song "Anthem" by Leonard Cohen: "There's a crack, a crack in everything / that's how the light gets in."

Freed of prearranged talking points, the dinner gave fifteen virtual strangers permission to be vulnerable, to engage as human beings in an open and genuine conversation, and to surprise one another and themselves—very different from the more formal interactions they would be having throughout the conference. No one knew what to expect when the evening began, and when we closed the dinner, no one knew exactly what had happened. Someone said afterward: "It was deep but not heavy." Another one wrote to us: "It's the first dinner I've ever been to where I went in not knowing anyone and came out feeling connected with every single person." Hearts, minds, and bellies were full as everyone returned to their hotel rooms.

Inspired by the success of our Abu Dhabi gathering, we decided to turn 15 Toasts into a series, and a few months later we began to host more dinners: "15 Toasts to Fear" in San Francisco; "15 Toasts to Dignity" in Capetown; and "15 Toasts to Transitions" in Geneva.[11] More and more leaders from different walks of life took part, and some of them subsequently set out to host their own 15 Toasts dinners. We had created a small community of diners to prove that the world might be a better place if power and vulnerability could clink glasses at the same table.

"The best things in life happen around a table," Priya likes to say. The experience of getting to know strangers over dinner, loosely moderated, but well designed and curated, can be the perfect catalyst for transformative experiences. Most great partnerships begin with the simple act of "breaking bread" together. In fact, an artist friend of mine told me that he never started a collaboration without first baking a loaf of bread for his future collaborators. While not all of us can hold ourselves to such high culinary standards, we can certainly be mindful of the meal before us on important occasions. Exquisite food intensifies exquisite emotions.

No one knows this better than software entrepreneur Chris Muscarella, the cofounder of the start-up Kitchensurfing. Muscarella has been working on social technology for more than fifteen years. He joked to me in conversation, "I do peer-to-peer shit." Although he had already had some success by starting a company called Mobile Commons, which provided a mobile platform to many large nonprofits and political campaigns, he found himself feeling burned out. In 2009, he took a sabbatical from the world of technology and started to explore cooking. This naturally led him into the restaurant culture, and in 2011, he and some friends opened a restaurant in Brooklyn called Rucola.

Muscarella's immersion into the food and restaurant world of New York, combined with his experience with software design, led naturally to a new kind of peer-to-peer software platform. In 2012, Kitchensurfing was born, a service that connects at-home diners with trained chefs, completely bypassing the middleman (aka the restaurant). "I wanted to have more of my own dinner parties, but I didn't have the time to forage and then do all the cooking," Muscarella told me. "We like to say that every kitchen is an ad hoc restaurant waiting to happen, and every chef is a punk rocker who just needs a fire source."

Kitchensurfing allows chefs to have a more intimate and meaningful relationship with their diners. This idea may well democratize the accessibility of private chefs and catered dinner parties, but what makes Muscarella's venture arguably more interesting, certainly more romantic, is his mission: to initiate and support great conversations around food, and to use the convenience of digital technologies to create authentic spaces for genuine human interactions.

Few settings are more conducive to a thoughtfully structured conversation than an intimate dinner. A notable example is the Death Over Dinner phenomenon, a grassroots movement that brings together Americans across the country in small-group dinner-table conversations about the end of life. "Let's have dinner and talk about death," the invitations read, making the implicit argument that people find it easier to digest such a heavy topic over food. "We put forward this myth that we don't want to talk about death, but I think we just haven't gotten the right invitations," said Michael Hebb, who had the idea for the series after speaking with health care practitioners about the crisis of end-of-life care in the United States.[12] The Death Over Dinner website provides various templates and some helpful tips to get the experience right. "Our hope," Hebb states in his TED talk, "is to spark the gentlest revolution imaginable."[13]

Whether it is name tags, airport interactions, workplace designs, e-mails, or dinner conversations, the biggest delights come from turning our day-to-day experiences on their head, "hacking them" gently, cracking them open for some sunshine and a breath of fresh air.

I encourage you to prioritize joy over optimization. Identify and implement the smallest possible feature—the size of a Band-Aid will do—to make an experience more intimate. Design with the naive eyes of a child and find genuine moments of surprise and delight. Don't forget to laugh at your own imperfections. We certainly have big problems to solve, but the Business Romantic recognizes that we also have a bounty of easy fixes that produce big wins. After all, the accumulation of countless moments of attachment ultimately adds up to a lifetime of joy.

CHAPTER 4

Be a Stranger

> Strangers are endearing because you don't know them yet.
>
> —DEJAN STOJANOVIC

At fourteen, I was so shy that I often avoided going to the most ordinary of places. Whenever I did dare an outing—to the grocery store, say—I would reach for packaged meat and cheese instead of heading to the deli to avoid speaking with anyone. I hated being shy and saw it as a crippling handicap. Though I was genuinely interested in other people, I often felt trapped by my own self-consciousness.

Fast-forward to my current professional life: eight years ago, shortly after starting my job at Frog Design, I was invited by our PR agency to the Financial Follies gala dinner in Midtown Manhattan. It was the annual see-and-be-seen extravaganza of the finance and media industry. Even then, well into my thirties, that same adolescent shyness was still lurking, and I arrived at the event feeling nervous and anxious. Although I imagined myself ending up in a corner alone, I knew I needed to defeat my fear of the situation. Instead of seeing the evening as a critical business function (the very word "function" sending shudders down my spine), I romantically envisioned it as a magical encounter with fascinating strangers. I wore the mask of the extrovert and watched myself performing in it. By the end of the party, I was surprised at how quickly the time had passed. This act of transformation had dismantled some of my self-consciousness. I met new people; I even made some friends.

Now I attend up to a dozen conferences a year. I lead meetings, host employee town halls, and organize dinners and parties. Over time, my romantic approach to social events has become second nature. It has allowed me to accept my own shyness as a legitimate response to the world, and that, in turn, has made me less shy. I have learned to tell my story in sixty seconds, a skill I honed while living in L.A., where everyone has perfected the art of the blitz elevator pitch. I have also learned to tell it in sixty minutes, a skill I acquired while living in Berlin, where everyone has not just a story but a history. Late bloomer that I am, I realized at some point that posing a random question at a networking event is often enough to initiate a connection, and that most people will welcome an expression of genuine interest. The truth is, we are all just waiting to be tapped on the shoulder.

In this way, I evolved from wallflower to social butterfly. Making new connections and chatting up strangers has become my default mode. (It would be interesting to examine how many people choose to turn their greatest anxiety into their profession, but that's a topic for another time.) Considering this, I sometimes long for the integrity and the intimacy—the splendid isolation—I felt back when I was shy. In some sense my shyness was a manifestation of my respect for the world around me. When I was younger, every successful social encounter was surprising, and every new friend was a wonder. In adult life, it's all too easy for these encounters to become routine. For Business Romantics, this won't do.

Business presents a multitude of platforms for exploring different aspects of the self, and Business Romantics revel in the opportunities to shape and shift personas: from shy to outgoing; soft-spoken to loud; acerbic to sweet; or guarded to openhearted. Others may accuse us of being shapeshifters—so be it! As Nietzsche once said, "You are always a different person." This sense of difference both in ourselves and in others leaves us constantly guessing, sitting on the edge of our seats. Business Romantics seek out such strangeness in all forms of socializing. Our first contact with a stranger, opening up and entrusting ourselves to another, is a moment of wonder. This is how it all begins.

Of course, such a view differs starkly from the perspective of the typical networker, a cross between a predator on the prowl and a street performer putting himself before the public for evaluation, and ultimately, validation. I once saw a guy at a reception approach every single person in the room with his business card in hand—his finger on the proverbial trigger. Instead of inviting conversation, he passed on his card, demanded one in return, and then excused himself. I wasn't sure whether to condemn him for his bluntness or to admire him for his efficiency. What's the point of cutting-to-the-chase if there is no plot?

From a transactional viewpoint, networking events are attractive precisely because they are so efficient. At a good networking event, you can meet more "relevant targets" within a couple of hours than you could possibly meet through "targeted outreach" over the course of a year. Networking events are also a venue for cultivating cachet, which seems to require a preponderance of VIPs and so-called thought leaders, the more famous the better. Moreover, hosting a social event provides a legitimate excuse to e-mail hundreds of people you don't know. You can make an impression even if no one shows up (which made me once consider sending out invitations to a "secret dinner" that would never take place . . .).

But these are the rational reasons to continue networking. We all know, from a logical perspective, why we are supposed to stand around a room trying to be heard over the din of dozens of other invitees vying for attention. Business Romantics, on the other hand, are seeking an experience for which the immediate rewards are less obvious.

At one networking dinner I attended, the places at the table were constantly rotated, bringing everybody in contact with everybody else throughout the meal. At the end of the evening, I found myself seated next to a woman who turned to me and said, "You're the only person here I haven't talked to yet!" We smiled at each other, and then both of us moved on, an unspoken bond cemented by what hadn't been said. I can no longer remember most of my conversations from the dinner, while the face of that woman remains an indelible memory. We formed a genuine connection without ever really having a conversation.

Nothing is more off-putting to the romantic than a whiff of the disingenuous. We are wary of networking events that are disguised as something else, called "dinner with friends" or "fun, informal get-together." Of course business is why we're all here, and yet none of us is really attending *strictly* for business. No one wants to be part of a herd. We attend because we want to catch some excitement or glimmer of hope; we seek out exchanges that might offer the possibility for a meaningful, even life-changing, conversation. We show up because we can't help but wonder, What if?

Take Harald Neidhardt, who shuttles between Hamburg, Germany, and San Francisco. A serial entrepreneur, he is the founder of MLove, a conference network that hosts its main annual event in an old and run-down castle near Berlin.[1] Each gathering brings together a small group of attendees from the worlds of business, academia, and the arts—from the Lufthansa IT manager to the Silicon Valley entrepreneur to the Romanian hacker—for three days of conversations that center on technology and society. The location is remote, in the East German hinterland, to avoid distractions, and it encourages participants to fully commit to the experience. Many MLove attendees, and even speakers, sleep in a makeshift dorm in the castle, among strangers and a far cry from the comfort and perks of other exclusive gatherings. Only one floor and one grand stairway separate pillow-talk and keynote. Initially, the *M* in MLove stood for "Mobile," but now, five years into his endeavor, Neidhardt says it has expanded to "Meaning." Inspired by the more intimate early days of the TED conferences and Burning Man—the legendary festival in the Nevada desert celebrating self-expression through the mechanisms of the barter economy—Neidhardt set out to create the quintessential conference for Business Romantics.

"I wanted to host a business gathering where you feel less alone," he told me. There are artistic installations, musical performances, social games, and dinners at MLove, and of course also talks. The focus is on creating moments of awe and revelation. Networking is beside the point, and the event never exudes the air of a typical conference. "If attendees bond, they will surely come up with a business idea that can help them stay connected," Neidhardt said, laughing. "My notion of success is to make people's eyes glow."

Gianfranco Chicco, a serial conference organizer who has curated numerous innovation and technology events in Europe, is even more romantic in his ambitions. He told me he wants to host a "conference for two" one day. It is sure to be the most exclusive conference ticket on the market. Try one at your office.

STRANGERS IN THE HOUSE

We find our greatest source of friction—and thrill—in encounters with strangers. Opposites attract each other, common wisdom holds. The most fruitful example of this opposition comes from the strangers in our own house: outsiders on the inside. These are the contrarians and rebels who question the status quo and deviate from the norm; they are the hidden gems of any innovative culture.

Businesses have found their own ways to institutionalize opposition. One is the "artist-in-residence"—a format used by various organizations, including IDEO,[2] Siemens,[3] and NASA, to allow individuals (artists, curators, scholars) to practice outside of their usual communities. There has even been an artist-in-residence program at the San Francisco garbage facility and recycling center since 1990.[4] In the best cases, the artist-in-residence plays the role of the familiar stranger. He is a corporate resident, but not a corporate citizen. Since he does not enjoy the same rights as the full members of the organization, he has more permission to experiment.

James George is Microsoft Research's first-ever artist-in-residence. Under the auspices of Microsoft's special creative and experimental unit, Studio 99, George spent three months creating coding applications that allow artists to engage with Microsoft's interfaces. "Artists approach coding from a different mentality and need a different language," George explained. "Microsoft was open to embracing the lessons we had learned from open-source initiatives and bringing that thinking into the platform itself."[5] By viewing their everyday tools through the eyes of a stranger, Microsoft engaged with an entirely different community of users.

In newsrooms, the so-called public editor operates in a similar role, although with a different mandate. Representing the interests of the readers and the broader public, he or she serves as an internal watchdog that holds the organization and staff to the highest ethical and professional standards, often probing editorial decisions and scrutinizing possible biases in coverage. Margaret Sullivan, the current public editor of the *New York Times*, for example, took her publication to task over what she criticized as insufficient reporting on the U.S. government's drone policy.[6]

Even the military, arguably the most hierarchical of human institutions, carves out institutional space for contrarians. The U.S. Army runs an internal disruption unit: the University of Foreign Military and Cultural Studies based in Fort Leavenworth, Kansas, also dubbed "Red Team University." The school's curriculum is designed to combat institutional groupthink in the military by empowering and training in-house skeptics.

The in-house skeptic, the public editor, and the artist-in-residence all raise serious questions about their host institutions and how they operate. Oscillating between friction and outright opposition, these are all characters with a high degree of independence, positioned away from the action but at the heart of the matter. They are the "misfits," the permanent strangers. They are the romantic hermits of today's business world.

Several initiatives are under way to give these outsiders on the inside a collective voice with a mobilized community around them. I introduced you to the League of Intrapreneurs, launched by Alexa Clay and Maggie De Pree, in chapter two. Clay and De Pree aim to equip contrarian employees with best practices and tools that enable them to self-organize more effectively. In addition, they aspire to raise executive-suite awareness of the potential of empowering such positive disobedience.

"One of the greatest difficulties social intrapreneurs face," De Pree told me, "is to maintain their outsider status. How do you strike the right degree of constructive discomfort to drive change from within?"

Corporate Rebels United[7] and Rebels at Work[8] are similar networks, and like the League of Intrapreneurs they seek to connect change makers inside corporations. Rebels at Work was originally inspired by a movement of internal "rebels" at the CIA that created the award-winning Intellipedia, a Wikipedia-like platform for sharing intelligence across government agencies. All three groups aim to provide the structural support needed to prevent corporate rebels from being pulled back into the normative behaviors of the larger culture.

Companies themselves have begun to embrace their in-house opposition and organize "hacks" or "hijacks" that challenge top-down initiatives and serve as powerful countermovements prompting a rethink or even a reset. Some have launched internal disruption units that can drive radical innovation from left field. Anheuser-Busch, for example, set up shop in Palo Alto with a "Beer Garage."[9] The outpost is the beer company's attempt to experiment with emerging trends in digital marketing under the radar of the company's more mainstream culture. McDonald's followed suit.[10]

As an alternative, companies may also bring in innovation consultancies—the likes of IDEO or Frog Design, which are essentially hired opposition—with the mandate to disrupt conventional thinking and overcome organizational myopia. The caveat here is that these outside interventions may lead to temporary changes but fail to become part of a company's cultural fabric for the long term.

Organizations that allow deviance and creative opposition accrue notable benefits: they lower the likelihood of "groupthink"; they show their employees that they value them as individuals and for their differences; they encourage an open exchange and therefore get a better feel for outside trends that might be emerging; and they stay ahead of the curve by anticipating potentially troubling mood shifts from within their organization. With a strong and self-organized in-house opposition, companies can cover the entire breadth of their corporate character. It allows them to acknowledge that they are complex and multipolar, that they have multiple truths inside them, and that, through this tension, they can become capable of stretching themselves, expanding, and realizing their full potential.

Business Romantics like to be hermits and misfits: strangers in their own house. But they also hold the space for other contrarians and rebels, even if these groups turn against them. They understand that those who disagree with the present often see the future more clearly, and they consider occasional disloyalty to be the strongest form of loyalty. So make sure your internal opponents have ample safe space to self-organize. Ask yourself: What is your "underground"? Who are your antagonists? Who is seeing the cracks in your organization and seeking to attack them? Invite them to do so before they invite themselves (and others along with them). Get them into the mix and not into the fold.

ANOTHER'S EYES

Our very expertise in our respective fields is often what blinds us to the most exciting and innovative solutions. The word "amateur" comes from the Latin word *amatorem*, which means "lover." Amateurs do things because they love doing them, not because they are good at them. They do not receive a paycheck or a performance review following their efforts, and yet they can give business some of its greatest inspiration.

LEGO, the iconic toy company, for example, was able to come back from near bankruptcy by connecting with the "lovers" of the brand. It refocused its energy on the LEGO fan communities around the world—involving both children and adults—and fully celebrated the creative work of its most devoted users. Some of these LEGO fans launched Brickfilms, a series of stop-motion animation films featuring characters made of LEGO bricks;[11] another "power user" created a photo series called "The Legographer," a collection of pictures taken daily on his iPhone that show the world through the eyes of a tiny LEGO man;[12] and yet another one created custom LEGO sets to ask his girlfriend, of course a LEGO lover as well, to marry him.[13]

At Frog Design, the most successful marketing campaign I was involved in was the work of an amateur, a "lover," who did not know what she was doing. Ashley Menger from our Austin office was not an amateur in her field; in fact, she was one of the best design researchers with whom I had the privilege to work. But she was an amateur in the sense that she stepped away from her profession and designed an experiment motivated by nothing more than the desire for personal inquiry. On her blog, she wrote one day:

> Who am I? I love disposable Jell-O pudding cups. I have designed five packaging schemes that are in full production. I recycle when I see a recycling bin. My landlord has a compost pile that I barely use. I feel a tinge of guilt, but really, it's not my problem yet . . . But that's not the person I want to be. I truly want to make a difference. From a high level, I have always felt that our massive garbage production is senseless and have never been afraid to say so. Yet I'm starting to feel the hypocrisy of espousing this green goal without living it on a personal level, on a daily basis. I have observed my own wasteful behaviors and have a lot of changes to make.[14]

So one day Ashley began to collect her garbage and carry it with her in a yellow plastic bag. By the fifth day, the bag had assumed considerable proportions, and she had become an unusual and much-talked-about sight on Austin's streets. Ashley took action to raise awareness of our (over-)consumption behaviors, and her drastic, hands-on measure was tangible and easily understood. It was a self-experiment that not only taught her about her environmental impact in the world but also how to deal with exposing her personal behavior publicly, including the mix of support, admiration, criticism, and scorn that it earned.

Ashley started a diary on her blog called *Trash Talk* in which she documented her daily garbage output and shared insights into her consumption patterns. Her initiative quickly became popular, and it did not take long before her colleagues at Frog joined in. Within days, we had garbage-carrying employees in Shanghai, Seattle, and Milan. A week later, after we had begun to prominently feature *Trash Talk* on our website and amplify the effort via our social media channels, followers outside of Frog launched their own *Trash Talk* experiments. The media took notice, and *Trash Talk* turned into a mini-movement.[15]

What made that happen? *Trash Talk* combined dedication to a good cause with personal activism. It was not just another cause-marketing program or a company funding laudable philanthropic activities. *Trash Talk* was launched by one individual with a desire to be held accountable. Ashley did it without following any particular rules or professional edicts. She was simply curious: an amateur is, above all else, a lover of learning. Doing things with the spirit of the amateur makes our eyes fresh: it allows us to experience and appreciate the strange in the familiar.

This tension between the strange and the familiar was also made manifest in one of the most remarkable marketing campaigns of recent years: Dove's "Real Beauty Sketches." The personal care brand hired an FBI-trained forensics artist—a master of the criminal sketches that often appear on "Wanted" signs and television reports—to draw two respective portraits of ten women it had selected. In the first, he used only their own descriptive language, asking them to answer the question "What do you look like?" but never once looking at them with his own eyes. The second portrait he created was based on a stranger's description. Different strangers were paired with each one of the women and given an opportunity to speak with them. These strangers then sat in the same room with the forensics artist and described what each woman looked like.

Dove released a documentary about this project that culminated in a final exhibition of each woman's portraits displayed side by side.[16] One by one, the women walked into the gallery and looked, first, at the drawing based on their own description, and then, second, at their portrait as seen through a stranger's eyes. In each pairing, the first portrait appears more severe, more closed off—less "beautiful"—than the second. It is hard not to be moved while watching the women come face-to-face with their diminished self-perceptions paired next to those of the world. "I have some work to do on myself," one woman said in disappointment after comparing her two portraits.

The Dove series used the eyes of an outsider to reveal the beauty inside each of the women profiled. Sometimes only a stranger's perspective can make us appreciate what has become all too familiar.

It calls to mind a story my wife once told me about one of her first jobs after college. She worked at an advertising firm in New York, where she reported to a vice president. Occasionally, on Fridays, she and her boss would change roles: they dubbed it "Freaky Friday." They would swap desks, take each other's respective calls, and even make some decisions on behalf of each other. Straight out of college, my wife had initially been hired as an administrative assistant, but she quickly became a jack-of-all-trades project manager. Her boss was a seasoned executive, and she took mentoring very seriously. Swapping roles was one of the many ways she enabled my wife to stretch her wings. But "Freaky Friday" was not just beneficial for the junior party in the equation; my wife's boss also left her comfort zone, and this ritual gave her an opportunity to understand her business through another pair of eyes. The role swap made Fridays different, and the unpredictability of the day was exciting for both of them.

Imagine if my wife's boss had simply stayed at her desk and allotted time on her calendar to learn more about other perspectives. As absurd as it sounds, we make this mistake all the time, treating empathy as an abstract concept that can be analyzed apart from our lived experience in the world. Empathy needs to be felt in the body, perhaps best of all through a physical act like walking. The German biologist Jakob von Uexküll once proclaimed: "The best way to find out that no two human world views are the same is to have yourself led through unknown territory by someone familiar with it. Your guide unerringly follows a path you cannot see."[17] The installation artist Janet Cardiff, famous for establishing audio and video walks as artistic formats, invites participants to experience the world in an entirely new way by moving through space while listening to her fictional narratives through earphones.[18]

Corporate training sessions have begun to model seminars on the work of artists such as Cardiff. They create exploratory thematic city walks as an avenue of inspiration for business leaders. Being forced off the beaten path, participants have an opportunity to see the world from a different perspective. In San Francisco, a group of corporate change makers, artists, and scientists regularly goes on "thinking hikes" in the Marin headlands, each one devoted to a specific topic. Such thinking hikes can be used to bridge the gap and mend tensions that arise within organizations. Companies could even institute regular walks for pairs of colleagues who, due to their function and role, tend to be antagonistic, or at least nonsympathetic.

One case that immediately comes to mind is the natural tension between the chief marketing officer (CMO) and the chief financial officer (CFO), a classic antagonism in the modern corporation because the agendas are typically so misaligned: the CMO likes to spend money to build a brand and win customers, whereas the CFO's mandate is to enforce fiscal responsibility, rein in expenses, and hold the spenders accountable for the return on their investments. The consequences of this misalliance can range from atmospheric disturbances to passive-aggressive feuds to outright war. The lack of mutual understanding is commonly due to the opposing mandates, not necessarily personality issues. During my tenure as CMO of Aricent, the engineering services firm, I had various disagreements with the CFO over budgets as well as the strategic direction of the company, but after he departed the company, he sent me a nice e-mail. He told me he had the utmost respect for my work, and that all his opposition was simply because he had to "put on the CFO hat."

If you can put on a hat, you can take it off as well. Instead of e-mailing each other, imagine the CMO and CFO going on a sixty-minute "thinking hike" every other week to discuss things. (Cc's might occur in the form of special invitations to select people to join.) This will feel strange at first, but imagine how effective this moment of shared strangeness will be for revealing yourself and recognizing the other. If you're the CMO, it might even increase your budget. And just think how much richer your professional life will be if it is a collection of walks, and not meetings.

WANDER WITHOUT MAPS

Every business plan is a map of sorts that pinpoints a company's position in the marketplace in relation to others. We need our maps, but as Business Romantics, we also need to reinvent them and even throw them away from time to time. The artistic and political movement Situationist International—mainly active in the sixties and seventies—played with what it called "psycho-geography."[19] In one of their most famous experiments, the founding members walked across a region of Germany with only a map of London to guide them.

Such an example might seem extreme, but we walk around with conflicting cultural maps all the time. Bring your colleagues together and ask them to draw out a floor plan of the offices. Each person will create a different drawing based on his or her experience of the space. Akin to the artistic experiments of the Situationists, give the programmer in your office the map drawn by your receptionist. What happens when one takes the other's journey? Ask your consumers to draw a mental map of the company strategy. Does it correspond in any way to the map drawn by your CEO? How does it look compared to the mental maps of your competitors' strategies?

As an alternative, try to wander through work on a different path. If walking with others—management-by-walking—is an act of empathy, forcing us to match each other's step in an attuned choreography, then a wander alone is an exercise in mindfulness. The flaneur, the famous archetype of the urban wanderer, leaves his apartment in the morning with no plan. He walks out his door with only a hat and a cane, open to mystery and serendipity.

New York City has neighborhoods designed for the flaneur as well as flat, gridlike neighborhoods designed for the man or woman of business to get transactions done. Walking along the grids allows one to go quickly from place to place, to see exactly what has come before and what lies ahead. Midtown Manhattan is a perfect example of this. Recently, I was able to make four or five meetings a day while never leaving a ten-block radius.

The West Village, on the other hand, is the flaneur's territory.

Filled with twists and turns, mysterious dead ends and idiosyncratic half streets and avenues, it is made for wandering. Although it is hard to get anywhere quickly, it is easy to find things to discover. The West Village is a very difficult place to do business, but it can be a lovely place to end up for an afternoon. How much of your own work space is designed, metaphorically, to resemble Midtown Manhattan? What if you cultivated a West Village in your office? Every once in a while the Business Romantic takes work away from the grid. Create obstacles to efficiency. Force people to look up and interact. Bring departments together for no other reason than to discover each other. Encourage your customers or employees to engage with your company as a flaneur. If they are guided, in a predictable way, from one thing to the next, give them corners to turn, space and story to uncover.

This type of discovery requires that enough informality is built into your organizational design. At Frog Design, a company that we employees amicably referred to as a "forty-year-old start-up," it was often said that in order to thrive "you need to be the water, not the rock," meaning you had to not only find but also *create* your own path through the organization. This demanded great nimbleness and flexibility, and, above all else, a surfeit of curiosity. But it only works as an organizational principle when there is enough open space, enough uncharted territory, to reward the wanderer.

When you think about it, this is one of the key design principles of innovation in general. Indeed, Bell Labs, arguably one of the twentieth century's greatest idea factories and the home of inventions ranging from the transistor to the laser to the cellular phone system, was designed with this very principle in mind. The president of Bell Labs, Mervin Kelly, believed in encouraging scientists of different disciplines to mingle. He purposely designed long hallways placing physicists next to chemists next to electrical engineers so that the researchers would be forced to bump into one another and begin conversations.[20]

Recent studies have found that an overlap in daily walking patterns indeed increases the level of collaboration.[21] Today's leading innovative companies such as Google, Samsung, Salesforce.com, and Tencent have all made serendipitous encounters an integral part of their workplace designs, and architects have developed intricate computational models that don't leave anything up to chance. For its Las Vegas headquarters, online retailer Zappos even introduced the new metric of "people collisions" (with both colleagues and strangers outside of its building) to foster creativity.[22]

However, these examples of carefully engineered serendipity cannot replace genuine unpredictability. Allison Arieff, who writes about architecture and design for the *New York Times*, commented: "The only people that you're really running into are your co-workers, and at some point that leads to a certain level of navel-gazing, I think, because you're only ever talking to people who sort of agree with what you're saying."[23]

This phenomenon applies to other aspects of our lives. Our "social graphs," a phrase coined by Facebook to describe our online and offline social ties, are in danger of overdictating our exposure to one another. As digital technology has become ever more sophisticated at personalizing and customizing our social experiences, online and off, each of us is increasingly stuck in our own cultural and moral fiefdoms—"Filter Bubbles," as Internet activist Eli Pariser, the board president of MoveOn.org and the founder of the media company Upworthy, has called them.[24] These are the work of algorithms programmed to constantly reflect back to us variations of our own image. The more we feed them our "likes," our "clicks," and our "choices," the more of ourselves the algorithms volley back. Soon, if we're not careful, we begin to live in a house of mirrors. All of our content is specifically designed to look like us. But who is not us? Where are the strangers? We have no idea.

Romantic encounters are at risk of being reduced to algorithms: matchmaking is now done by dating sites; community is curated less by a sense of citizenship than by shared "preferences"; and the serendipity of travel is being minimized by exclusive features such as private security services, frequent flier privileges, and social travel applications that let travelers even pick their seatmates on the plane.[25] The result is more options but fewer opportunities for random encounters. We have more meetings without ever really meeting anybody.

Technology can aggravate the pain, but it can also be the cure. "20 Day Stranger," an iPhone app jointly developed by the MIT's Media Lab and Dalai Lama Center for Ethics and Transformative Values, promises "a tiny window onto a wider world." It allows two strangers to share their experience of the world—anonymously—over the course of twenty days, and foster empathy and tolerance in the process.[26]

The rise of the sharing economy—business models that enable peer-to-peer resource sharing among consumers—also has the potential to expose us to more strangers—and strangeness. Consider Airbnb, the online marketplace for private vacation rentals or bed-and-breakfast accommodations. Without doubt, the service is disruptive.[27] Just a few years ago the idea of renting out your private home to total strangers would have been completely unfathomable, but Airbnb has made it a widely accepted business and social practice. There are evident economic reasons for Airbnb's success, but for romantics the company's main value is its ability to bring back a quality of strangeness to travel. Airbnb users share a small piece of their lives with each other. Sure, they are are given choices, and they typically enter an agreement with ample knowledge about the other party, but ultimately it is still a gamble for both sides, and serendipity remains part of the game. You just never know exactly whom you're going to meet—and that is the very romantic thrill of it.

Other examples of collaborative consumption services offer similar benefits, beyond merely being a utility: think of ride-sharing providers such as Carpooling, Uber, or Lyft. All these companies operate on the basis of networks that facilitate ad hoc on-demand connections between drivers and passengers. Users rate drivers, drivers rate users, users rate users; and the resulting trust scores minimize the concerns when getting into a stranger's car. But here, too, an element of taking a chance remains, and the thrill of the unknown is a key ingredient of the experience. In fact, aside from gas savings, the social element—the possibility of meeting strangers—is a big part of the success formula. Carpooling, for example, boasts on its site that it has led to "more than 16 marriages, and thousands of friendships."[28] It is interesting to note that in one of the original rideshares, the casual carpool from Berkeley over the Bay Bridge to San Francisco, strangers picked each other up and made up weird rules for what could or could not happen in the car.

Uber, the controversial "Transportation Network Company," as it describes itself, epitomizes the romantic potential of the share economy. Because of the density of Uber's network and the on-demand nature of its service, you are more likely to quickly connect with an Uber driver in your vicinity than to successfully hail a cab. When you request a ride through the Uber mobile app, the interface gives you the name of the driver and also his or her head shot, service ratings, current location, and car model. Then you have a few minutes—usually less than five in a city—to anticipate the stranger being sent your way. When he arrives, you can greet him by name—"Hi, George"—forging a little bond between the two of you right away. The transactional aspect of the relationship is invisibly tucked away with account credit cards on Uber's web back end so that both rider and driver are free to simply experience the evolving conversation. Uber's business model is more convenient and efficient than that of taxis, but the Business Romantic values it for other reasons. A taxi provides a service; Uber provides a serviceable stranger, one who kindly offers you a ride.

By capitalizing on excess capacity, the pioneers of the share economy have made consumption more collaborative and more convenient. The next, more romantic step might be to also make us *happier* by tapping the excess *social* capacity: all the strangers who are just one word away from making a difference to our lives—and theirs. A study conducted by the behavioral scientists Nicholas Epley and Juliana Schroeder illustrates the potential of random micro-interactions.[29] The two researchers asked commuters to chat up strangers on a train to work and then asked them about their experience, comparing it with that of commuters who experienced the commute without any socializing. Those who engaged in casual social interactions reported overall more positive emotions. Even the research subjects themselves were surprised by this finding, as they had predicted the opposite and would therefore typically refrain from making contact. This gulf between our expectations and our actual experiences indicates that we consistently underestimate the importance of small moments of attachment, as well as the impact random strangers can have on our happiness. Weak ties apparently matter at least as much as strong ones.

The researchers Michael Norton and Elizabeth Dunn, authors of the book *Happy Money*, more recently ran a related experiment, asking Starbucks customers to have a "genuine conversation with the cashier" rather than just sticking with minimal, efficient interactions.[30] Again, those who socialized and put human connection over efficiency, generous with their time and attention, reported being more cheerful afterward.

SELLING

Salespeople are perhaps the most transactional people in business, or at least you would think so given their role and performance metrics. There is only one, rather brutal measurement that counts: how much money you bring in. There is little ambiguity, and nothing to be argued with; if the targets aren't hit, heads roll. But salespeople are not only the most transactional people in business; they are also some of the strangest. They are often outsiders within their industries, and often even outsiders within their organizations: "sellers" who are disconnected from the "doers."

Salespeople—like any other cohort in business—all have different styles. There are "farmers" (those who grow existing accounts); "hunters" (those who go after new prospects); and "closers" (those who can bring the deal home). And yet, the job seems to attract a certain type of individual with a universal set of attributes: flexible, aggressive, hungry, bullish, and eternally optimistic. The salespeople with whom I have worked have all been warm, friendly, and gregarious—and then, suddenly, they were gone. "Didn't meet his target," I was told when I asked the human resources department about what happened to one of my former colleagues. That's why salespeople are sometimes considered the "mercenaries of business": given the volatile nature of their employment, they tend to move from gig to gig, shifting their loyalty and identity to different organizations quickly—or perhaps not identifying with anything at all.

In my job, I regularly get sales calls from outbound sales reps who want to sell me conference sponsorships or CRM (customer relationship management) solutions (there seem to be many of them). One guy once left a six-minute sales pitch in three separate voicemail messages. "Getting a lot of no's will get you to getting more yeses," the sales 101 wisdom has it, but, still, I wondered how in the world cold calls like this, especially on the grounds of painfully obvious superficial profiling, could possibly be effective.

I should know. I was once a cold caller myself, as a part-time telemarketer during my undergraduate studies. Working three afternoon or evening shifts a week, I was part of a call center staff, handling calls in response to TV commercials that advertised everything from medals to stamps to pay-TV subscriptions. I both received calls and made calls through a sophisticated technology platform that was designed to route every call to the right agent. While I was expected to "upsell" orders on the inbound calls, the true sales efforts were the outbound calls, where I—armed with a step-by-step script and countless FAQs to guide me through virtually every possible interaction scenario—was asked to cold-call strangers of all ages and backgrounds, sometimes at ungodly hours, "hard-selling" them all kinds of items. One night I called more than fifty women over fifty in an attempt to sell some of them, or even just one, anti-cellulite tights.

I had mixed feelings about the job: on the one hand I was appalled by the telemarketer's (my own!) chutzpah to bother people with unsolicited offers at their homes. I was as intrusive as a salesman at their doorstep, if not worse. On the other hand, I secretly enjoyed learning firsthand about the psychology of selling. Moreover, the experience of sitting in a call center—which more closely resembled a call *factory* with hundreds of small terminals and agents—was an exercise in cognitive dissonance: vibrant and exciting, but also bizarre and alienating. There were times when I felt like a member of a herd of cattle, deprived of any human dignity, and then there were moments when I felt most alive as a human being, thriving in connectedness to "the other" in the most direct and honest way imaginable.

We are all salespeople: some of us persuade others to buy something; some of us persuade others to buy *into* something. If markets are sympathetic communities for social exchange, then selling is the search for a *sympathetic individual.*

The most astonishing aspect of my short telemarketing career was how intimate the conversations were with some of my "targets"—some of them so lonely that they were grateful for the opportunity to talk to another human being (senior citizens would often launch into monologues about their grandchildren, making it next to impossible for me to hang up). Business often begins when the romantic connection ends, yes, but business can also provide a last resort for any kind of connection at all. Like receiving a message in a bottle from a far-off shore, a business transaction can provide genuine human contact in a sea of isolation.

Selling gives us the permission to relate to the other without having to engage in the full emotional implications of a relationship. Anyone who sells anything offers a piece of himself, and perhaps the Achilles heel of salespeople is in fact their very unique selling proposition: when rejection is your daily bread, you don't fear rejection so much anymore. At the core, being in sales is deeply romantic. It requires a perpetual state of unfulfillment: the salesperson always longs to touch ever more targets, to reach once-inconceivable new heights. It is a journey without end. The numbers may add up from time to time, but they will never be enough.

The sociologist Georg Simmel described the stranger as someone who "comes today and stays tomorrow."[31] He is in the group, but not *of* the group; he is more distant than he is near. This stranger is a common figure in novels, theater, and film, but he also has an essential role to play in business. If we truly intend to "do business together"—what psychologist Steven Pinker defines as the ultimate gesture of peace[32]—we need to get to know each other, not better, but differently. So keep wondering: Who are you? Act like an amateur. Resist conformity and give misfits a place in your house. View the world through another's eyes and wander without maps. Take a chance. Take a ride. Make one thousand cold calls for one moment of human warmth. (Un)expect the kindness of strangers. Not everyone can be known. Not everything can be familiar. Isn't that cause for celebration?

CHAPTER 5

Give More Than You Take

I know what I have given you . . . I do not know what you have received.

—ANTONIO PORCHIA

Every December, for more than twenty years, Tom Taylor and Jerome "Jerry" Goldstein, a San Francisco couple, have transformed the exterior of their gingerbread Victorian in the city's Noe Valley district into a phantasmagoric display of holiday cheer. "Tom and Jerry's House," as it is known to locals, sits atop one of the city's steepest streets and features a giant lit-up sixty-foot Christmas tree (a Norfolk Island pine purchased by the couple in 1973 as a houseplant) as well as a bounty of colored ornaments and bears. Above the garage door (repurposed as a fireplace) hang two eight-foot-high stockings replete with stuffed animals. An oversize toy train, including a polar bear, tootles around a K'nex Ferris wheel, and one spends at least fifteen minutes counting the myriad animatronic teddy bears embedded in the carefully curated tableau. Santa makes an appearance every evening, often accompanied by hosts Tom and Jerry handing out pamphlets and presents. Over the years, the house has become a popular destination for kids and their parents, a citywide attraction eliciting wonder and delight and more than a twinge of civic pride. It even inspired a documentary: *Making Christmas: The View from the Tom and Jerry Christmas Tree.*[1] Tom and Jerry do not accept donations, for the tree and their house are a "recognition of the spirit of the holidays, inclusive of rich and poor, young and old, children of all ages, and the beauty and spirit and diversity that San Francisco offers."[2] Put simply: Tom and Jerry's House is a gift.

In his seminal work on arts and the economy, *The Gift*, poet and cultural critic Lewis Hyde posits that gifts, unlike commodities, create relationships.[3] By giving someone a present, we recognize him or her, and this acknowledgment is the greatest gift of all. Gifting also involves an element of surprise: who would ever think to ask for Tom and Jerry's House? And yet countless families have been truly delighted by it. A good gift does not satisfy our immediate needs; it reveals our latent desires. It widens our horizons and stretches our souls. Gift cards, on the other hand, are the ultimate antigift: they delegate the choice back to the recipient. The gift of money—inherently fungible—reduces the gift exchange to a mere transaction.

The magic of gifts calls to mind a story I once heard about the filming of the movie *The Deer Hunter.* While the crew was shooting the famous wedding sequence, director Michael Cimino encouraged the many extras to treat the festivities as a real wedding, so as to increase the authenticity of the scenes. Prior to filming the wedding reception, Cimino instructed the extras to take empty boxes from home and wrap them as if they were wrapping real wedding gifts and bring them to the set the next day. The fake gifts would then be used as props for the wedding reception. The extras did as they were told; however, when Cimino inspected the props he noticed that the "gifts" were a lot heavier than empty boxes otherwise would be. He tore the wrapping paper off a few of the packages, only to find that the extras had in fact wrapped real gifts for the wedding.

When you watch the film today, it's hard not to imagine the pleasure and delight in the minds of each and every one of the actors in the room, the joy they must have felt at handling real gifts, chosen by real people. Isn't it always slightly disappointing at Christmas, for example, when department stores put up displays with boxes and boxes of fake gifts under the trees? In this case, gifts are mere decoration like holly or ivy, with no inherent essence or meaning apart from their surface appearance. But, as *The Deer Hunter* story shows us, fake gifts are meaningless. There is something spiritual that emanates from a genuine gift box, something that speaks to the spirit of the people at both ends of the relationship.

Business is beginning to embed gifting and generosity in its transactional framework. "Create more value than you capture," recommends the technology consultant and Internet publisher Tim O'Reilly with regard to products and services that build sustainable ecosystems.[4] In his book *Give and Take*, Wharton business school professor Adam Grant explores the benefits of generosity at work.[5] Grant posits that workplace altruism is an undervalued source of motivation. Companies should have a strong interest in fostering giving behavior as it enhances key aspects of their performance, including effective collaboration, innovation, service excellence, and quality assurance. One of Grant's studies, for instance, suggests that interrupting employees' work by giving them occasional altruistic tasks increases their sense of overall productivity. Likewise, Grant says, a willingness to help others is at the heart of a fulfilling career. He cites the advice of Adam Rifkin, a serial entrepreneur and the most connected person on social network LinkedIn: "You should be willing to give up five minutes of your time for anybody."

Being generous with one's time may well be the greatest gift in business. For more than ten years, Frog Design hosted the opening party of South By Southwest (SXSW), the influential annual gathering of the technology community in Austin.[6] But the term "hosting" doesn't really do it justice: Frog came up with a different theme every year (from "Playing with the Time of Light" to "The Other Singularity"), designed the event experience, and built all kinds of interactive artifacts and installations to be displayed on site. What had started as a small event for friends and the local community grew over the years into a must-attend party at SXSW and one of the biggest of the tech industry overall. In 2013, more than three thousand conference-goers attended.

While working at Frog, I had the privilege of witnessing the passion of the Frog team in Austin that put it all together year after year, led by the tireless Jared Ficklin, the design genius and master of ceremonies. Whether it was real-world versions of computer games such as Electro Tennis and Human Tetris, hybrid creatures such as Zen Robots, or, my personal highlight, motion-sensor-enabled Port-A-Potties (with video projections that displayed when a potty was occupied and if the person using it was sitting or standing): if it was weird, Jared and crew could build it.

The most remarkable thing about the party, however, was its extracurricular nature: the Frog team worked on it mostly in its spare time: on weekends and during countless night shifts. Despite this, the opportunity cost was high. A still-significant amount of billable hours went into the production of the event, and the line between passion and obsession was routinely crossed. On strict business terms, I wish I could tell you that the investment yielded significant returns. Surely the party garnered nationwide media coverage, and it gave us the perfect opportunity to impress and hang out with clients, sometimes even leading to new project work. But frankly, this stood in no comparison to the hours the Frog team invested and how much more effectively we could have spent the money on targeted business development efforts. The party took over almost our entire Austin studio in the three months before the slated event. Every year we asked ourselves the same question: "Is it worth it?" And the answer was always the same: "Of course!" It became a no-brainer. The party would happen. It was nonnegotiable.

Business Romantics recognize the beauty in all of this effort. The fact that we had no quantifiable ROI for all our commitment and creativity transformed the party from a product into a precious gift for hosts and attendees alike: it was the result of an excessive dedication, a curiosity for the strange, and a longing for the unbounded.

Business offers other, more unexpected avenues for gifting. On April 1, 2013, Google announced a beta version of "Google Nose," acknowledging that for years the company had neglected the sense of smell as a critical device for search.[7] Quickly revealed as an April Fools' joke, the concept spread rapidly throughout the Internet. As every marketer knows, the first of April presents a significant branding opportunity, especially for brands that have well-established expectations they can break.[8] But something else is interesting about the April Fools' jokes of brands: they are a quirky example of generosity in business. They slyly give more than they take. Of course clever marketers primarily use them to drive brand impressions and loyalty. But, more important, they also give us permission "to be a fool." In that sense, Google Nose and all the other April Fools' jokes make the world a little bit more unreasonable, a little bit more romantic, just for one day.

Gifting opens up opportunities for alternative types of currency in our everyday lives as well. One morning in my own local coffee shop, a man treated every single customer in the shop—

close to a dozen—to an unlimited amount of coffee. The barista seemed slightly embarrassed when he announced the gift, but everyone in the room happily accepted, thanking the generous spender. Once everybody had their coffee, the man started a little speech, saying that he had bought everyone coffee to get a couple of minutes of our attention. He then pitched his nonprofit, essentially "buying" our time with his gift.

A few weeks later, I was at lunch with a friend at a restaurant in downtown San Francisco. My friend and I were so immersed in conversation that I didn't even notice that my wallet had been stolen until I reached for my credit card to pay the waiter. We were sitting in plain daylight on an outdoor patio with only a few other patrons. I felt suddenly off-kilter with this new crack in the veneer open; I was vulnerable. A stolen wallet is like a missing key in your keyboard or a sudden ache in your tooth—it jolts you momentarily out of your placid assumptions about the world.

My friend kindly lent me some money so I could get my car out of the garage and drive home. I immediately called the credit-card company and had my cards blocked, and then I went to bed with a severe headache induced by the mere thought of all the IDs that needed to be replaced. The next day, the doorbell rang, and an unfamiliar male voice asked: "Tim?" I expected the worst, and still grumpy, barked back, "Yes. What do you want?"—assuming a sales pitch. The man said: "I have your wallet." I walked down the stairs to the door. The stranger, accompanied by his young boy, claimed he had found the wallet and identified me through my driver's license. All my cash was gone, but my credit cards and my driver's license were still there. I was immensely relieved, but also a bit flabbergasted. How should I respond? What was the man's motive? Was it mere kindness? I thanked him profusely and then asked him, "May I give you some money?" although I wasn't sure if this was really the right question to ask. His answer was a resounding "Yes." He pointed out that it had taken him some extra gas to drive to my house. So I handed him a twenty-dollar bill. He thanked me; I thanked him again—and then he left me standing in the doorway, confused by the whole interaction-turned-transaction.

Both instances—the "coffee bribe" and the "returned wallet fee"—initially gave me a sense of disequilibrium. These acts of kindness had hidden transactions inside them. At first I felt a bit cheated or duped and kept thinking that when kindness becomes a currency it stops being kindness. Upon reflection, however, I changed my mind about both events. The man in my coffee shop had bought my attention. And the stranger who returned my wallet may have done it in the hope of a reward. But both of them created meaningful moments for me, and I was touched by the *experience*, the *perception* of generosity in both instances. For romantics, motives are ultimately incidental. Giving is there for the taking.

The use of kindness as a currency is also on display in the Suspended Coffee movement.[9] The idea is simple: people pay in advance for a coffee meant for someone else. For example, they order and pay for four coffees and tell their barista that three of them are "suspended." The customer then takes only one for him- or herself, and the barista can give the others away. This exchange embeds a secret gift inside conventional transactions. The idea of Suspended Coffee reportedly started in Naples, Italy, before becoming a viral hit on Facebook. It has since spread all over the world and grown from coffee to sandwiches and even whole meals.[10]

Reddit, a social news and entertainment website that aggregates user content, has initiated its own hybrid form of gift and transaction. Random Acts of Pizza, or RAOP, allows Reddit users to buy pizza for fellow users based on how compelling they find their "pizza requests."[11] Before submitting a request, of course, the pizza-hungry must become Reddit users. And, after receiving an RAOP, it is assumed that the user will "Pizza-It-Forward," buying pizza for someone else. Users connect through their wit and wordplay. The whole system thrives not because the pizza gifts are so overwhelmingly generous, but because the process is fun.

The chocolate brand Anthon Berg opened a pop-up store in Copenhagen that operated in a similar fashion: the "Generous Store" asked customers to pay for their chocolates not with money but with the promise of a good deed for a loved one.[12] Like Suspended Coffee and RAOP, it turned a transaction into an interaction and converted generosity into a "hard" currency.

Outdoor clothing company L.L. Bean, famous for its absurdly generous return policy, is often a victim of its greatest gift. Stories of twelve-year-old backpacks and smelly boots landing in the return pile are legion. What does L.L. Bean say, each and every time? "It was a pleasure doing business with you."

Social media offers the perhaps most powerful arena for generosity these days. On Facebook, Twitter, Instagram, or Tumblr, we constantly give more than we take. We post into the ether, often without response, without immediate return. We overshare. According to a 2013 study by the Pew Research Center, almost one-fifth of adult Internet users have posted personal videos; many hoping, says Pew, that "their creations go viral."[13] And this quick sharing of events and content can trigger waves of exponential giving.

Liba Rubenstein, the director of strategy and outreach at the microblogging service Tumblr, sees a new kind of social activism emerging from this culture of sharing on social media. During the 2012 presidential debates, for example, when Mitt Romney made his ill-fated remark about "binders full of women," Tumblr was one of several social media sites that almost instantly took the phrase viral.

"It crystallized what was really at stake in the election for a lot of women," she told me. "And then people took the idea offline:

they were adding their own interpretations to it. I saw women dressing up as 'Binders Full of Women' at real Halloween parties."

For Rubenstein, this example became emblematic of the fluidity between online and offline engagement. "It often starts as a joke, but it hits a nerve because people feel strongly about it. They take action offline, and then they contribute that back into the online conversation," she said.

Political and social movements inspired by viral clips are often leaderless at the beginning. It is impossible to track down the very first "share," and, in the midst of a viral sensation, it is also irrelevant. Like a catalytic flame, viral moments burn bright and then die. The triggers can be profoundly trivial or truly profound. On one end of the spectrum, there is the pop-cultural residue such as the Little Darth Vader commercials,[14] LOL cat videos,[15] or the *Downfall* movie parodies.[16] On the other end, you have historical events, from the celebratory to the catastrophic, causing social media mass attention.

At its cynical low, this collective sharing can be viewed as a tit-for-tat marketplace of egos desperate for recognition, and at its worst it can even lead to a mob of online rage, riot, and roar that everyone must follow. For every successful Kickstarter campaign, there is a "shit storm" somewhere else on the web. Think of the "HasJustineLanded" hashtag campaign, for example, when PR executive Justine Sacco's controversial tweet about HIV/AIDS in Africa ignited a real-time witch hunt on Twitter.[17]

At its best, however, social media sharing can result in stories expressing our very humanity. The Make-A-Wish Foundation created a viral sensation with its "Batkid for a Day" campaign that brought the dream of a five-year-old boy suffering from leukemia to life: to be a superhero for a day, cheered on by thousands of volunteers in San Francisco and a global online audience.[18]

The "First Kiss" video is another example of such viral love. Featuring ten pairs of strangers who were asked to kiss each other for the first time, it became an Internet sensation in early 2014 and garnered more than 50 million views within just a week after it was posted. The video shows the clumsy conversation as the strangers try to break the ice and ease themselves into a forced act of intimacy. Some of them move around each other awkwardly first; others are openly flirtatious and can't seem to wait to start the action. Then, to the tune of the song "We Might Be Dead Tomorrow," the couples kiss for the first time as if it were their last. Most of the encounters are pretty intense affairs, with the kissers visibly—and very much to their own surprise, it seems—overwhelmed by the power of the experience. It is fascinating to observe how a brief moment of intimacy with a stranger outweighs the lightness of the staged occasion.

After "First Kiss" became an Internet sensation, however, some negative reactions quickly followed suit. Commentators on the web accused the video of being fake, pointing out that it was commissioned by a clothing company and that the kissers were professional models.[19] "First Kiss" was indeed an ad for the Los Angeles–based boutique fashion label Wren, whose logo was displayed in the opening and closing titles. The budget for the video was only about $1,300 (mainly for studio rental), and none of the kissing strangers, all friends of either the label owner or director, were paid.[20] Be that as it may, regardless of the motives, "First Kiss" provided millions of people online with genuine moments of affection. Like the Coffee Bribe or the Stolen Wallet, it contained a beautiful interaction within a transaction.

Or consider the story of NPR host Scott Simon, who live-tweeted his mother's death to his more than a million followers in a series of tender and loving tweets, sharing, in intimate detail, how the two of them spent her final hours together in the ICU. The tweets drew an outpouring of support and condolences, and led perfect strangers to tell Simon that he had made them burst into tears and think about "good deaths and good lives."[21]

In these instances, viral sharing brings us together in collective mourning. When public figures such as Steve Jobs, Nelson Mandela, actor Philip Seymour Hoffman, or writer Maya Angelou pass away, these sad events lead to "peer-to-peer grieving," as writers Paul Ford and Matt Buchanan call it: "Real-time chronology, trending subjects, and curated news feeds mean that the Internet, with its mix of individual expression and automated sorting, writes the first draft of the eulogy."[22]

On the day of Steve Jobs's death, October 5, 2011, millions all over the world mourned the loss, and the overwhelming universal grief underscored that the Apple cofounder and CEO was truly one of the last business "tycoons." In both his work and his death, Jobs elicited our most profound sentiments, making him a subject of near-religious adoration. Almost everyone believed that without Jobs—without his vision, leadership, and aura—Apple would become just another excellent tech company. The romance was gone.

On the day Steve Jobs died, I was working at Frog Design. Jobs had engaged the firm in a seminal collaboration on the first Apple computers in the eighties, and as a result of it, Frog rose to fame. In the following decades, upon every launch of a new Apple product, we had made sure that the gadget and technology blogs would feature our glory days working for Apple, and we never tired of pulling out some old concept or prototype from our dusty archives for an easy PR win. Our collaboration with Jobs was the gift that kept giving.

After hearing about his death, I remember how we instantly decided to dedicate our entire home page to a farewell message that simply stated: "Thanks for everything, Steve." We even blackened out the navigation so that for seventy-two hours no one could get anywhere deeper on the site, let alone contact anybody or buy our services. For three days, our website basically went out of business in a final salute to Jobs. Initially, there was some concern about removing the top navigation, too. After all, even Apple's website, which displayed a photo portrait of Jobs and a link to an obituary, still featured links to other landing pages; the show had to go on. But a junior member on my team, with an infallible instinct for the mood of this moment, insisted on a more radical, truly "purist" gesture, and so we went with it.

The decision wasn't the result of careful deliberation or smart PR calculus. It just felt like the right thing to do. Our gesture touched a nerve, and colleagues, partners, clients, and even competitors told us afterward they had been moved by our homage. We probably lost some business leads, and we probably generated tons of goodwill instead, but most important: we stared at the page, again and again. It was our moment to give.

Business Romantics make generosity—in fact, an obsessive kind of generosity—the default strategy. Helping others fosters our emotive engagement with the world and lets us connect with something greater than ourselves, even if it's just our neighbor. The easiest way to subvert the reciprocity of markets is to simply give more than we take. From Tom and Jerry's House to April Fools' jokes to Frog's SXSW party to the Generous Store to viral love and grieving online: our actions are romantic when they expect nothing in return.

Make sure you invest in what you feel strongly about. Be excessive in your intention, attention, and execution. Overpromise, overcommit, overdeliver. When we strive for beauty, pleasure is often the collateral. After all, some people are more likely to commit random acts of kindness after eating an amazing piece of chocolate. The romantic wouldn't have it any other way.

CHAPTER 6

Suffer (A Little)

> But I don't want comfort. I want God, I want poetry, I want real danger, I want freedom, I want goodness. I want sin.
>
> —THE SAVAGE IN ALDOUS HUXLEY'S *Brave New World*

Tartine bakery in San Francisco's Mission district is a popular destination for coffee and pastries. It is so popular that a long line of hungry customers serpentines down the street every morning starting at seven A.M., well before the shop opens. The bakery is said to have the best croissants outside of France. I drive by Tartine every day on my way to work, and I am always fascinated by that queue of the well-heeled waiting so patiently. Having grown up in Germany, I am reminded of the socialist society of the former GDR, where lines in front of shops were a common sight as people queued up for basic necessities (hardly a luxury item like

Tartine's pastries). I still have a hard time reconciling these two images. After all, aren't choice and convenience supposed to be the insignia of a market economy? Why wait in such a long line when there are plenty of coffee shops and bakeries along neighboring streets without any lines at all?

My guess is that people like to wait for three reasons: The obvious one is that the pastry at Tartine is so exceptionally good that it is worth the wait. (Personally, I think it's not, but hey, I grew up in Germany, so I prefer pretzels over croissants.) The second reason is that the line affords people a public space for meeting friends and neighbors. But it is the third reason that is most significant for romantics: waiting in line represents a little bit of suffering; it is an investment of time and effort in return for some exclusivity. The waiting at Tartine is part of going to Tartine; it is a key ingredient of the experience, a key ingredient of the brand. Tartine has become a popular destination precisely because it requires so much effort to arrive there. You need to *earn* it.

Common wisdom holds that convenient customer experiences create stronger customer loyalty. Romantics know this is a misconception. Customers of bakeries with no lines might be more satisfied, they might even be happier, but their experience will be lacking something when compared to a vision quest like Tartine. We recognize elements of this phenomenon in all aspects of our consumer lives. Camping out for days in front of an Apple store for the release of a new iPhone, waiting in line for hours at Disneyland to spend three minutes with Mickey Mouse, or strenuously hiking up to the mountain plateaus of the Alps to catch a few seconds of the leading bikers flying by during the Tour de France: in all of these cases, the very effort dramatizes the moment and increases the perceived value, the exclusivity of the experience.

Take customer loyalty programs such as frequent-flier miles. Part of their operating principle is to delay gratification, to force customers to be patient in return for eventually getting their needs fulfilled. It is remarkable that we will waste our time on anything in an age where instant shipping and near-real-time delivery have become the norm. It is also striking that, in such cases, we are able to take the long view—typically not one of our strong suits, as every psychologist will tell you—rather than thinking short term. This is because customer loyalty programs make us feel special from the moment we sign up for them. The reward is all the more meaningful to us because we had to invest in it—not just pay for it, but *earn* it through our loyalty.

IKEA, the Swedish furniture retailer with lean-and-mean self-assembly products, has perfected the art of frustrating its customers. IKEA doles out its experience of pain into two parts: first, the shopping parkour, which, with its standardized layout and one-directional traffic, sends customers through a seemingly endless loop of cheap and chic Scandinavian-style rooms before releasing them into the limbo of flat-pack boxes on ten-foot-high shelves. ("Oh, no, sorry, we don't have that in stock anymore.") If it feels like the store itself is the seventh circle of hell, the misery continues at home with the self-assembly. There is often a screw that doesn't fit, a synapse that's missing, or the ultimate moment of defeat when the newly self-assembled model collapses after hours of labor. Putting together a Billy, Oslo, or Klippan is an exercise in sacrifice, a painful reminder of our own existential incompetence. Given its unique model of suffering, it is no surprise that the IKEA experience has become a cultural meme with its own community folklore, including a blog proclaiming the "10 Commandments of IKEA Furniture"[1] (e.g. "Thou shalt take thine time") or a web quiz called "IKEA or Death" that tests your knowledge of IKEA product names by comparing them to the names of death-metal bands.[2]

A recent study published in the *Journal of Consumer Psychology* coined the term "IKEA-Effect" to describe the phenomenon that consumers tend to place a higher value on products they put together themselves.[3] The researchers concede that this requires completing assembly of the product; otherwise outright annoyance is sure to eclipse the sense of ownership. They argue that the right level of effort is critical: too little effort and consumers don't emotionally invest; too much effort and they are put off by the inconvenience. Either way, effort is an integral part of the experience.

All these examples of customer experiences tell us that we overrate the importance of convenience and underestimate the role of sacrifice. What is rewarded is our commitment, and this commitment gets reinforced with every reward we are given—it is a virtuous cycle. The more effort we accumulate, the more committed we become. And frustration is a big part of the equation. "Labor Leads to Love," as the title of the IKEA study suggests—and there is no love without suffering.

This is precisely why we appreciate conferences and events held in inconvenient places. It is not easy to get to the annual meeting of the World Economic Forum, held in the Swiss mountain village of Davos (unless you are among the lucky ones who are flown in by helicopter or private jet). Once you do get there, you are often met by freezing wind and icy sidewalks. And yet it is never suggested that the World Economic Forum move to an airport hotel near Geneva. Why doesn't the Camino de Santiago in Spain have an express train? Couldn't the Burning Man Festival take place in the lovely wine country of Napa Valley instead of the Nevada desert? There is a reason that we, as a people, have created a history of pilgrimages to our most meaningful—and spectacularly inconvenient—gathering places.

Evolutionary psychologist Ulrik Lyngs studies the relationship between life challenges and happiness. When I met him in London, he described the mismatch between our biological apparatus and our evolved external conditions as the conundrum of "Stone Age brains in the Space Age." In his view, we are still wired to compete for survival, but in most modern, civilized societies existential threats have all but disappeared. In a seminal article on "The Evolution of Happiness," Lyngs's colleague, the psychologist David M. Buss, refers to the "large discrepancies between modern and ancestral environments" as our main impediments to achieving happiness in modern life.[4] Because we live in an environment that is significantly safer and more stable than that of our ancestors, we lack the "critical events" that help us distinguish between false and real friends, assess our fitness, and give us any sense of achievement. This lack of critical assessment events, Buss writes, may cause "the loneliness and sense of alienation that many feel in modern living, a lack of feeling of deep social connections despite the presence of many seemingly warm and friendly interactions." Business Romantics have a more succinct way of putting it: we lack drama.

This elimination of drama is due, in some part, to the advances of technology. Software has made us softer. It has made our lives safer and more certain, but also more domesticated. In the age of digital commerce, instant gratification and total convenience have become the default modes of interaction. We can now order pretty much everything online and have it delivered to our doorstep. We even outsource our relationships to services such as Delightful, a "personalized date concierge"; HowAboutWe, which curates romantic experiences for married couples; or mobile messaging services such as BroApp or Romantimatic, which automate intimate communication.[5] As if all that weren't enough, BreakUpText even sets up prompts for automated breakup notes. The creators of the app first conceived of it as a joke, but not everyone found it funny: "This app isn't very good, my breakup texts are always a lot better," one user complained.[6] And the pinnacle of it all is Wevorce, an online service that helps couples "uncouple" and outsource their divorce process.[7]

In an era of global competition, for both brands and individuals, convenience, ease of use, and instant consumption are the lowest common denominators, the basic requirements for all products and services. But with everything outsourced and automated, simply made *too* easy, where do we find those "critical events" to punctuate our lives? Some turn to skydiving, mountain climbing, and ultramarathons, while others attempt to set up IKEA furniture and wait for hours in line at a lauded bakery.

At the workplace, on a day-to-day level, many of us use deadlines to give us the thrill of suffering. Deadlines are our equivalent of scaling Mount Everest. Remember the PowerPoint deck for your boss? How you hadn't even started on it the morning of the day it was due; how the "end of business day" deadline seemed far away until it suddenly got dark outside and your coworkers left the office? You bought a latte from the café around the corner, went back to your desk, and stayed there until dawn, hammering out the slides until your eyes began to ache. When you thought you were done, having finished the final lap, you realized that you still needed another hour to write the accompanying e-mail to your boss. Finally, you logged off, turned off the lights, and took a taxi home, where you crawled into bed, tired but happy.

Staying up late, working extra hours to get a proposal or presentation done, is an emphatic experience that heightens our awareness and challenges our physical limitations. We both dread and relish the experience. The work—so easy to off-load—looms before our very eyes.[8] And yet the excitement of the challenge is also palpable. With our hearts racing, we thrill at the sensation of time getting thicker with every minute counting down. In the face of such professional near-death experiences, we come alive.

Architects and designers suffer from this last-minute-rush syndrome until late in their careers. The root cause lies in their professional education. As students, they are conditioned into this nail-biter race-against-the-clock mind-set early on in their training. It is no different for writers, brand strategists, advertisers, filmmakers, researchers, analysts, journalists, and other knowledge workers: once they've experienced the adrenaline rush, they have a hard time letting go of it, despite all attempts to eliminate panic mode and tame the volatility in the creative process. We secretly want drama and suspense, and we have developed an impressive ability to create it through our work.

Having deadlines, it should be noted, is a privilege for those of us who have a certain degree of autonomy over our work; those of us who are not chained to monotonous shifts or a standardized sequence of tasks. Being able to determine *when* we work, or at least *how* we work, increases our chances for creating drama through deadlines. As knowledge workers, we can intensify our experiences because we can determine when they begin and end, where they peak and where they pause. Without this freedom to procrastinate, romantic suffering is much harder to achieve.

This type of self-imposed sacrifice occurs all the time in the workplace, but it is most apparent among sports fans. It is no coincidence that the word "passion" comes from *pathe*, the Greek verb "to suffer." A fan is committed to his or her team and suffers through every defeat. The level of suffering is proportional to the level of devotion. The emotions fans express through sports events are akin to the experience of a personal crisis: from excitement and euphoria to shock, trauma, and even depression. And the most memorable moments are not the ones of joy but of exquisite pain: the heartbreak after defeats.

To me, the most romantic sport is soccer, because the "beautiful game," as it is often called, offers not only drama but also a high degree of ambiguity. "The great fallacy is that [soccer] is first and last about winning," Robert Dennis "Danny" Blanchflower, the legendary captain of British team Tottenham Hotspurs, once said.[9] To prove his point, it is part of the rules that soccer games can end in a draw, and even goalless. Arsène Wenger, the coach of Arsenal London, went so far as to say: "The imprint you make on the spirit of the people is more important than the result."[10] In the best case, fans are rewarded with stunning, seamless combinations of ball passing and ball-free movement that lead to almost divine geometries of beauty. Top-performing teams such as FC Barcelona or Bayern Munich elevate the merely mechanical, the merely physical, to a sublime pleasure of enormous poetry: they turn running into ballet.

Despite the recent trend to fetishize statistics and the introduction of video evidence to overrule referee decisions, soccer has maintained its subjectivity, its unpredictability. Anything can happen, and almost anything that does is due to human error. When performance is not rewarded with the proper result, when the "soccer god," in fan lingo, does not do justice, the fan's heart bleeds, and frustration mounts. But the fan stays devoted, and the team's history—the victories and seminal defeats—becomes *his* or *her* history as well. Game after game, season after season, these memories create layers, one atop another, until a fan's life is lived in tandem with the team. Even in the face of the most frustrating and painful losing streak—such as the trophy-less "drought" of the legendary Brazilian club Corinthians between 1954 and 1977[11]—the fan would never switch loyalties and look to support a more successful team: being a fan means the faith was established long ago. As in all of our most meaningful relationships, this emotional investment will never fully pay out. It will absorb and exhaust us, and there will be no exit—but our romantic capital will grow with every game we watch.

Moments in business can give us a sense of significance because we invest valuable resources such as time, care, or attention. The more inconvenient the experience, the more significant it feels. The more we sacrifice, the more we belong. The closer the deadline, the more we come alive. There is nothing more terrifying to a romantic than a life of complete ease and convenience: all the time, anytime, anywhere. Is that what you are offering to your customers and employees? Why?

Business Romantics are dying to suffer (a little). We know that in the cracks between comfort, satisfaction, and happiness romance awaits us. Ask your customers and colleagues to make an effort. Make it more inconvenient for them to achieve gratification. Let them try harder. Let them wait. Frustrate them so they value what they cannot get. And reap the ultimate reward: permanent unfulfillment. The desire intact, the thrill not gone.

CHAPTER 7

Fake It!

> Everything is false, everything is possible,
> everything is doubtful.
>
> —GUY DE MAUPASSANT

The Australian director Baz Luhrmann's movie *The Great Gatsby* was a box-office hit when it was released in 2013. It was also considered a total fraud. Film critics railed against Luhrmann for putting his auteur stamp on the world. Instead of looking like a "real" depiction of Fitzgerald's "real" novel, critics argued, Luhrmann created a fantasyland, a carnival ride of the garish and the grotesque.[1] Unlike many reviewers, I loved the film. I felt as though I was looking at Gatsby's world of East Egg through a disco globe. Everything had a shimmer to it; the world was realer than real. In fact, it was a celebration of the fake.

Luhrmann is no stranger to this aesthetic—we have seen him work the same cinematic magic on nineteenth-century Paris with

Moulin Rouge and Shakespearean Verona in *Romeo + Juliet*. Anyone looking for a faithful rendering of Fitzgerald's universe would do better to stay home and read the book. But why would we want it any other way? After all, every work of art is attempting to render—to fake—an artist's experience of reality. As Picasso famously put it, "I often paint fakes."

Consider the graffiti artist Banksy, a provocateur the *Guardian* dubbed "a master of the obvious."[2] Using stencil paintings and epigrams to subvert the facades of existing buildings and objects, Banksy delivers his artistic vision with a healthy dose of civil disobedience. He is even more famous for his headline-making stunts, constructing false appearances and false notes in an attempt to reveal truths. He once left an inflatable doll dressed as a Guantánamo prisoner in Disneyland, and another time he employed a real shoeshine man to polish the shoes of a traveling Ronald McDonald statue in New York.[3]

A similar kind of provocative fakery was at play in an awareness campaign UNICEF ran in Finland. "Be a Mom for a Moment" placed blue strollers with a crying baby audio track in crowded places in fourteen cities. When people looked in the strollers, they found a note with the message: "Thank you for caring, we hope there are more people like you. UNICEF: Be a mom for a moment." The public reaction was overwhelming, with coverage in all major TV, radio, and web news. The estimated media reach was more than 80 percent of the Finnish population after two days.[4]

The artifice of Banksy's work and the UNICEF campaign resonates with us romantics because we naturally prefer intrigue over information, emotion over evidence. A line in Donna Tartt's Pulitzer Prize–winning novel *The Goldfinch* masterfully articulates the alchemy between truth and illusion:

> Between "reality" on the one hand, and the point where the mind strikes reality, there's a middle zone, a rainbow

> edge where beauty comes into being, where two very different surfaces mingle and blur to provide what life does not: and this is the space where all art exists, and all magic.[5]

As Business Romantics, our sensitivity to this magic allows us to celebrate the fake while dismantling universally held truths. We thrive in this constantly shifting space, using it as a playing field for provocation and subversive commentary.

Nowhere is this more evident than on the social web, where the line between the real and the fake is quickly disappearing, leaving a no-man's-land for tricksters and provocateurs to exploit.

Twitter is a prime example. As Karen Wickre, one of our Business Romantics from chapter two, can attest, the public nature of Twitter allows us to perform, play, try on poses, and stage our own perspective on the world. The Twitterverse is filled with false personas and pranksters: accounts such as @WillMcAvoyACN and @PaulRyanGosling parody mainstream culture by providing an instant feedback loop to society's inanities. A Twitter feed such as @MayorEmanuel takes it even further by creating an entire account commenting on the travails and gaffes of Chicago's mayor Rahm Emanuel. These Twitter personas, and countless others, remain completely anonymous—masked for years—while attracting hordes of followers.

Or take Eric Jarosinski: in the real world, he is known as a scholar of German literature and philosophy, but his fans all over the world know him by his Twitter persona, @NeinQuarterly. Several times a day, Jarosinski's loyal followers get a taste of his unique romantic sensibility in the form of aphorisms, wordplay, and raw missives from everyday urban life ("Kapital walks into a bar. Leaves with two."). Combining dark sarcasm with disarming candor, Jarosinski uses philosophy and literature to expose the quirks and biases of our hyperconnected lives. "I was never drawn to artifice," he told me, "but this feels like such a natural way to connect. Twitter is actually the place where I can find the freedom for all the ideas I am interested in: irreverence, playfulness, provocation."[6]

Fake campaigns used by activist groups invite us to fully embrace this ambivalence in the world of brands. The "Shell: Arctic Ready" online ad series lampooned the oil giant's push to drill in the northern shores of the United States (one ad displayed the sinking *Titanic* along with the tag line "Never Again: Let's Go!").[7] Burger King's hacked Twitter account announced that the company was sold to McDonald's and that all their employees were on drugs.[8]

This melding of the fake and the real was also illustrated in burrito and taco chain Chipotle's ad campaign "Scarecrow," featuring an animated short set to the sound of Fiona Apple's gleeful cover of "Pure Imagination."[9] In keeping with the brand's mission to "Cultivate a Better World," the ad vilifies factory farming through a dystopian vision of imperiled chickens in high-rise prisons. It didn't take long before comedy artists released a parody called "Honest Scarecrow," which mocks the original video, changing the lyrics of the song to "pure manipulation" by a "giant corporation."[10] The remarkable thing is that both films, the original and the parody, are effective and surprisingly moving. The art, the beauty, and the charm of the original, even though brilliantly called out as a cynical fake by the satirical version, survive the creative mockery.

These same experiences of authenticity amid fakery have a place in our lives as consumers and workers as well. As Business Romantics, we revel in wearing masks. From Grimm's fairy tales, to the carnival in Venice, to Mardi Gras, Halloween, Batman, and hacker network Anonymous—masks allow us to escape not only others but ourselves. They magically transform us from individuals into icons. Religions and cults have long understood this, and they use masks as symbols of humanity's most archetypal roles. In this way, masks rest at the fault lines between the literal and the symbolic, the utilitarian and the transcendent. They are worn for physical protection—think of fencing, oxygen masks, or the antipollution masks we have grown accustomed to seeing in Asia—but also for purposes of disguise and performance.

As consumers, we put on masks and try on different identities as we enter the showrooms, as we buy and buy into the products, services, cultures, and values of the brands we revere. Ralph Lauren, a boy from the Bronx, for example, famously invited his customers to try on the mask of the tony and well heeled.

We all wear masks at the workplace, too. We perform by completing tasks and accomplishing goals set mostly by others. But we also enact our own narrative by choreographing our interactions and playing different social roles. These types of performance have become ever more essential to our "performance review." The knowledge economy has automated many of our quantifiable, concrete tasks and left us with only the fuzzy space of *subjective* tasks: shaping perceptions; building and cultivating relationships; managing our reputation; curating and sharing tacit knowledge; earning respect, popularity, authority, and influence. As Matthew B. Crawford claims in his book *Shop Class as Soulcraft*, we have become "symbolic knowledge workers":[11]

> A manager has to make many decisions for which he is accountable. Unlike an entrepreneur with his own business, however, his decisions can be reversed at any time by someone higher up the food chain (and there is always someone higher up the food chain). It's important for your career that these reversals not look like defeats, and more generally you have to spend a lot of time managing what others think of you.

If one were to grossly exaggerate, one could say we are no longer what we make or do—we are who others think we are.

We are how much we are liked or feared. We are what we are *not* told. This has far-reaching implications. Our projects and initiatives involve more and more departments, more and more stakeholders, and as a result, we have drifted further and further away from any satisfying sense of our own work's completion. As we do away with the linear, cause-and-effect "to-do" list, the only thing left for us to do is "to be": to build and live up to the promise of our personal brand, serve as the symbols of our work, and show-and-tell the world about it.

That's highly volatile and vulnerable terrain, and many of us need more than just one persona to navigate it. Thus we wear masks while we are at work, and we perform rituals that give a shape to our fragile identities: think of all the weekly, monthly, and quarterly reports; the recurring check-ins and reviews; the daily stand-up meetings common among product development teams; and the offsite retreats. Think of the special committees, task forces, and councils that are created not just to disperse power more widely (or the perception thereof) but also to provide us with multiple stages on which to perform. And think of the parallel growth tracks clever human resources teams have invented to give our various identities more space for actualization, often distinguishing between "title" (e.g., senior associate, principal) and "role" (e.g., retail practice leader, regional marketing manager). Some knowledge workers might view all of these rituals as a morass of corporate bureaucracy, but Business Romantics appreciate each and every one of them as opportunities to explore different sides of their selves. The hierarchies may be flatter now, but there are more avenues than ever for our symbolic knowledge work, in other words: for faking our performance.

This desire for expansiveness, for a story of our work selves that bleeds outside the lines of a linear narrative, is finally being reflected in social media's networking tools.

One start-up venture, aptly named Somewhere, is poised to

replace LinkedIn and more conventional résumés. A little bit like Pinterest for professionals, Somewhere is a visual discovery platform for the enterprise. Individuals or teams post images and captions to tell the stories of their work, and by doing so, they can share what they love doing and how they do it ("What do you really do?" the site asks poignantly). Somewhere allows users to present themselves as contradictory and complex human beings, their passions, emotions, and aesthetic preferences all on display. These narratives connect them with other, like-minded professionals along the way. When you browse through Somewhere's site, it's immediately evident that the presentations of the "working self" are aesthetic creations. Everyone is faking it! For romantics, this is a wonderful thing. Careers are finally presented as ambiguous and ever evolving, and the professional's identity is rendered as a fluid persona. Justin McMurray, cofounder of Somewhere, told me that he created the company in 2012 to put workers "back at the heart of and in control of telling the stories of their work. People lead such fascinating work lives: our hope is that we can help open up the world of work, help people see behind the scenes, find inspiration, and find the people they should be working with."

Of course, not all work worlds are so immediately fascinating or aesthetically cohesive. In 2005, George Nachtrieb, a film and TV producer as well as a performance artist, was producing content for Amgen, the biogenetics firm, when he was struck with an artistic impulse.

"We were in that stereotypical office with no windows and rows and rows of cubicles," he told me. "During meetings, it was a totally accepted convention for people to get up and simply stand in front of their PowerPoints and read their slides word for word. No one ever said, 'This is a terrible way to convey information.'"

That experience, along with a long history of artistic experimentation, led Nachtrieb to create an alter ego, Steve Musselman, and an entirely fake corporation: YD Industries (YDI), producers of everything from contraceptive dinosaur sponges to "Butter Pups," a push-up snack for kids, to tofu-fed beef. Musselman regularly presents seminars at performance spaces in Los Angeles and San Francisco, exploring topics such as "How to Fail" and "YDI Success," including the requisite breakout and brainstorming sessions.

All of this might sound more like satire intended for artistic circles except that Nachtrieb's work is gaining popularity with actual businesses. Living in San Francisco, the heart of start up culture, he realized that the seminars' tightrope walk between truth and artifice provided a fruitful playing space for exploring the innovation process. He even recently teamed up with a former producer from design firm IDEO to create seminars for start-ups.

"We did a session for a group of engineers, and they were completely confused," Nachtrieb said, laughing. "But soon into the presentation, almost everyone eventually caught on. It gave them a chance to see their own practice with some perspective. Engineers are very focused on what is possible, and they tend to work within a rigid practice."

Nachtrieb's presentation taught them theatrical improvisation techniques such as "yes and," a rule stating that every suggestion from an improv player must be met with a positive form of agreement from the other players. The alternate playing space of YDI—its absurdity and sense of wry playfulness—kept the engineers from dashing imaginative possibilities with practicality. After one of his recent public performances, a woman approached Nachtrieb and said, "I'm still not sure if any of that was real or not." Nothing could have brought him greater happiness.

The world of business should take inspiration from endeavors such as YDI and Somewhere and embrace the moniker "I" as a playful container of countless truths and lies, performances and rehearsals, offices and homes, on and off the clock. Stowe Boyd, a digital economy observer, points out: "In the new way of working, work isn't a place you go, it's a thing you do. It is you."[12] We are all faking it in order to feel it.

While employees are encouraged to embrace various truths, we don't give our leaders the same permission to be inconsistent. We stubbornly reject leaders who wear masks to exhibit multiple personas and switch between different facets of their character. In fact, unmasking has become the most recent business *cri de coeur.* In a blog post for the *Harvard Business Review*, Peter Fulda demands "Leaders, Drop Your Masks," exhorting them to "Just be yourself."[13]

The extreme scrutiny of the Internet audience has even further narrowed the range of appropriate expressions for business leaders. They now sit in a glass house where any hint of emotion can potentially cause a scandal. How can they respond to their impulses, act irrationally, or express an unforeseen emotion when thousands, or even millions, of eyes hold them accountable all the time? They are expected to be consistent, integrated, and predictable. Anything less than that and they are accused of "flip-flopping," or worse, labeled a phony. What we project and how we perform must always match. That's our baseline for integrity.

But one life is not enough. As consumers, employees, or leaders, the desire to escape uniformity and consistency drives us to borrow traits from masters of illusion such as the impostor and the con artist. Con artists and entrepreneurs, in particular, share many similar characteristics: the con artist sells a vision knowing it will never materialize, while the entrepreneur likes to believe it will.[14] Both of them know that getting fooled is all too human. "The big secret about art is that no one wants it to be true," as psychotherapist Adam Phillips put it.[15]

The "impostor syndrome" at work is widespread and well documented,[16] and the best way to counter it is to celebrate it.

Harvard psychologist Amy Cuddy, who studies nonverbal behavior in its relation to power, proclaims, "Fake it 'til you become it."[17] She illustrates her point by sharing her own story: as a young Harvard professor, she felt like an impostor, doubting her capabilities as an academic. Then she realized that her feeling like an impostor might be the very remedy: She began to fake a posture of confidence and competence in those early classroom settings. In the beginning, like an actor in early rehearsals for a role, her posturing felt awkward. Over the years, however, as she grew more comfortable, the role started to meld seamlessly with her sense of herself.[18] Pretending to be somebody else is the first step to becoming somebody else. When we act like impostors, what we really do is stretch our wings, creating room for more expansive versions of ourselves.

Don't get me wrong: I do not advocate that Business Romantics become con artists or masters of deception. I do, however, recommend that we acknowledge that reason often has little to do with our commitment to a company or brand. It would be foolish to ignore that we want to believe in something we don't fully understand. Hope and faith are the engines behind most of our human enterprises.

No one knows this better than Americans. A nation of immigrants—and I am one of them—has a special appreciation for those who make it by faking it, at least initially. America is called "the land of promise" for a reason; a powerful imagination and an expanded notion of reality are key ingredients of the American spirit. We have always been more interested in the future than the past, and we tend to be irrationally exuberant about our projections: anything is possible, anything can become true if we just believe it will. We have Wall Street, Silicon Valley, and Hollywood, not to mention the daily theatrical spectacle that takes place "inside the Beltway." We dwell in hero myths and urban folklore. We have a soft spot for the great persuaders. From earliest childhood, we learn to tell our stories; how to shape the narrative of our lives, our script, and our personal brand. We are obsessed with our celebrities and love the story of the story, the iconography, the metanarrative, the making-of and horse-race angles. Of course it is a fine line between vision and hubris, miracle and mirage, leap of faith and fake, but even cynics must admit: the American Dream remains the most powerful romantic idea for anyone wanting to be in business.

As Business Romantics, we need not be afraid of fakery. To overcome uniformity and consistency, we can make truths from fiction; we can make beauty from illusions; and we can make transformative changes from within ourselves. Romantics are more interested in subjective experience than in objective reality, in the perception of authenticity over actual truth. Which one of the Chipotle ads was more moving: the real or the fake? The answer, for Business Romantics, is: both. We seek out this ambiguity in all aspects of our lives: the double, the object and its shadow, the mask and its wearer. This gives us the freedom to distort reality, to fabricate events, to pretend to be somebody else. We keep producing new illusions to protect us from becoming disillusioned.

As brand marketers, we can create fake brands and campaigns that lay bare real emotions. As leaders, we can wear masks to help us reveal the full scope of our personality: who we really are. As symbolic knowledge workers, we can render our narratives with a concern for aesthetics. As entrepreneurs, we can emulate the traits of con artists and envision a world that doesn't match reality, at least not yet. All of these efforts are born from the romantic desire to flirt with the artful lie, with the possibility of another life, with the idea that not everything is what it seems to be.

CHAPTER 8

Keep the Mystique

A smile is a door half-open and half-closed.

—JENNIFER EGAN

James Bond, the world's most famous spy, is one of our culture's best-kept secrets. Charismatic and ever opaque, Bond is never truly knowable. He pulls us in with magnetic attraction and then pushes us away with his aloofness and detachment. After a spate of more debonair Bonds such as Roger Moore and Pierce Brosnan, today's Bond the British actor Daniel Craig brings the iconic spy back to his romantic roots, closer to the original character, as conceived by author Ian Fleming. In his book *The James Bond Dossier*, Kingsley Amis describes the founding father of all Bonds: "He is lonely, melancholy, of fine natural physique which has become in some way ravaged, of similarly fine but ravaged countenance, dark and brooding in expression, of a cold or cynical veneer, above all 'enigmatic,' in possession of a sinister secret."[1]

Business can take inspiration from Bond's aura of secrecy. Don't tell us everything; leave us guessing. Secrets, when used judiciously, create possibility and suspense. In the midst of our current culture's obsession with transparency, they can make experiences feel thrillingly out of reach.

Of course, there can be a dark side to secrecy. When companies are fear driven and rich with paranoia, secrets can oppress and corrupt. I'm not promoting fear-inducing meetings behind closed doors, stealth projects behind the backs of colleagues, or a complete lack of transparency in decision making and consumer information. Clearly, secret fellowships among the elite can take on a nefarious tenor, revealing the fault lines between the powerful and the powerless. But don't disregard the value of secrets or secret meetings altogether. In their more innocent forms, they engage our empathy, our curiosity, and force us to consider what we hold most dear. When everything is open, nothing is open. What do we choose to close?

AMBIGUITY (OR NOT)

In business, brand is the designated domain for this kind of mystique. It is the intangible (and often invisible) asset, a source of enormous value for a business, but notoriously hard to grasp, let alone measure. Not surprisingly, the marketing industry has struggled to develop a universal methodology for quantifying "brand value" (for example, two leading rankings recently came to very disparate valuations of Apple: BrandZ says Apple's brand is worth $185 billion, while Interbrand says it's "only" worth $98 billion. The difference is a whopping $87 billion!).[2] Brand remains esoteric and opaque, and yet it is still the most powerful conceptual framework in business, a way of understanding and naming a collective identity based on myriad individual and shared experiences over time. It's a company's soft power.[3]

The concept of brand has significantly changed over the past two decades. As the effectiveness of mass advertising began its decline in the nineties, new marketing formats arose to address the increasingly dispersed and fluid nature of markets and fragmented audiences. The *Cluetrain Manifesto*, written in 2000, was a clairvoyant charter for the marketing profession. It foresaw many of the fundamental changes triggered by the rise of social media and ushered in the notion of "conversational marketing."[4]

Today, brands are expected to be content-rich conversationalists: they initiate, join, or curate online dialogues with interesting things to say, and they put forward opinions, points of view, and even arguments. In this new age, celerity is as important as adaptability: zig before you zag, flip before you flop. While the brand's values and its argument should stay consistent, experts recommend that the message evolve: because merely repeating the same statement again and again means you end up with a mind-numbing monologue—a commodity, not a conversation.[5]

As a consequence, brand marketers see themselves as reporters who follow and piggyback on stories that trend on social media. Some call this "newsroom marketing."[6] And with "native advertising"—another buzzword—marketers deliberately blur the borders to journalistic content. As many companies, including Cisco, General Electric, IBM, and Starbucks, have begun to create their own quality content, the lines between traditional media and branding are becoming fuzzy. Some well-respected journalists have already switched sides by joining firms with titles such as "content strategist" or "editor in chief." The brand megaphone has turned into a chatterbox, an ersatz friend weighing in on the national or global conversation—always accessible, personal, and transparent.

In the midst of this, consumers have become adept at identifying media and marketing spin, even if it's oh-so-clever, and they are now smart and informed enough to call a lemon a lemon, and a failure a #fail. "Dell Hell," blogger Jeff Jarvis's online rants about Dell's products and customer service,[7] and Canadian musician Dave Carroll's 2009 YouTube hit "United [Airlines] Breaks Guitars"[8] are classics.[9]

JPMorgan Chase experienced its true brand perception firsthand when it hosted a Twitter Q&A session in the spirit of openness and ended up enduring six painful hours of mostly harsh criticism and harassment ("@jpmorgan where do I send my resume? I'm smart and have very flexible morals").[10] And British entertainment retailer HMV even lost complete control of its social media team when some rogue members used the account to live-tweet a massive firing at the company (using the hashtag "hmvXFactorFiring").[11] Considering these, and many other examples, it is understandable if some brands lose their online countenance: Nokia New Zealand famously sent a simple "f--k you" to its Twitter followers.[12] Even friends have their limits.

Still, businesses need to accept the fact that their statements and their behaviors are now being monitored and scrutinized by a global audience that never sleeps. What happens in Vegas ends up on YouTube. But it's complicated. On the one hand, brands have less control because "your brand is what other people say about you when you're not in the room," as the adage goes. On the other hand, they have more control: thanks to social media, they can be in that room now, anytime. Proponents of radical transparency argue that a brand needs to stay connected 24/7, always "on," always in the conversation. But as transparency and openness have become the norm, they are now so inflationary that they no longer offer the appeal of genuine trustworthiness, personality, or intimacy. The pressure of maximum visibility and the demand for authenticity make it more daunting for brands to stand out and create meaningful connections to their publics.

This is where Business Romantics step in. Trying to reply to every single piece of information conveyed about us in the digital landscape is not our idea of a compelling conversation. Instead, like Greta Garbo, we withdraw—ever in demand, ever out of reach. We aren't interested in transparency as a virtue per se. A brand is, by design, more ambiguous, and it starts losing its power when it becomes too uniform, too consistent. In the age of conversations, it can be eloquent, but it must never be fully explicit.

Opportunities for re-mystification abound. Take the "brand book," for starters. Many brand strategists advocate creating brand guidelines with detailed instructions on the appropriate messaging, language, presentation style, design, and behavior to support the brand promise and persona. I have seen brand books as thin as two pages and others that were two hundred pages thick, with prescribed rules for every single brand expression you could possibly imagine—from the placement of images in PowerPoint decks to the language you should use for "elevator pitches" at networking events to the color of your tie. I have been responsible for creating several brand books in my career, and it took me a few years to realize why I would always feel an aversion to them. Of course I didn't like being regulated—I suppose no "brand ambassador" does—but more important, I didn't believe in the alleged purpose of a brand book. If it can be codified, if you can engineer it, it's not a brand.

The strongest, most distinctive, and most resilient brands are the ones that don't need a brand book; the implicit brand promise is understood (or, better, *guessed*) by everyone inside and outside of the organization. At Frog Design, for example, the company's DNA was so unique that new employees were not handed a brand book; rather, they were invited to a client meeting. "Jump right in and learn to swim in the pond," the motto went. Like ideologies, cults, tribes, and other intentional communities, brands cultivate a secret at their center, a mystical void that must not be filled. Apple may have a brand book, but the appeal comes from the personal aura that Steve Jobs embodied, and the orchestrated secrecy and heightened mystery he created around each new product launch. Collectively, we have loved and hated Apple—often feeling both extremes at exactly the same instant. This is what a great brand does to us.

The same applies to elevator pitches. I never really understood why people get so hung up on them, and I have outright refused to craft any at the companies I worked at. At Frog Design, we used to joke: "Elevator pitch? Good luck! You will need a long elevator ride!" Frog was a complex organization that required a great deal of tacit knowledge; a dumbed-down, standardized hundred-word blurb would never have been able to capture what "we were all about."

My sales colleagues at Frog wanted me to tout our work with blue-chip clients as much as possible, but many of those clients were reluctant to serve as public references (because of the not-invented-here syndrome, they were not keen on giving an outside agency any credit for their innovations). We had no choice but to turn this weakness into a strength, and over the years I came to genuinely appreciate the merits of doing so. I encouraged my colleagues to deliberately build up the mystique whenever they were asked about our client portfolio: "Well, it's tricky, you know, our work is so out there, so cutting edge, that most of our clients don't want us to talk about it. It will be on the shelves in two years or so, and then we may be able to talk about it." . . . "We work for almost all Fortune 500 brands, and chances are you have been using a product we designed." . . . "No, sorry, unfortunately, that's all we can tell you about it."

For Business Romantics, the most effective brand book and the best elevator pitch is the open-ended statement that leaves more questions open than answered. It doesn't provide clarity; it piques curiosity, and sometimes it even confuses. Notice how people lean in when they are given just enough information, which is to say, not quite enough: "Can you just explain that a bit more?"

A master of this type of mystique is McKinsey, "the most well-known, most secretive, most high-priced, most prestigious, most consistently successful, most envied, most trusted, most disliked management consulting firm on earth." So stated *Fortune* magazine back in 1993.[13] More than twenty years later, the statement still rings true, which is a testament to the brand's enduring strength after it pioneered the profession of consulting services to senior management. McKinsey, often simply referred to as "the Firm," was the first to establish a powerful mix of superior analytical smarts and generalist management strategy that claimed to trump in-house expertise and specialist industry experience.

But there has always been more to it. "You can't get fired for hiring McKinsey," the saying goes, and it illustrates the emotive factors that are an integral part of the company's success. With the outcome of strategy advice often hard to extrapolate, these factors weigh even more strongly than in other industries. Investment banker and merger specialist Felix G. Rohatyn poignantly remarked: "How can you judge advice? You should ask the people who got advice to tell you how they feel."[14]

The viability of professional consulting services is predicated on the perception that the consultant knows more than the client. Management consultants, in particular, are "solution shops," as the management thinker Clayton Christensen calls them.[15] They enter a client organization from the outside with the mandate to disrupt the status quo. They prescribe a "solution" and, by proxy, give their clients the legitimacy to make (often unpopular) decisions. McKinsey, in particular, perhaps more so than its peers, enjoys an invisible contract with its clients. At the core of its business model rests a highly intangible asset: an *assumption* of authority that endows *actual* authority. The Firm's recommendation remains unquestioned because it comes with an elitist aura infused with a powerful sense of mystery.

Other professional services—such as architecture—have boxed themselves into a world of time-and-expenses pricing based on cost that rarely reflects the full scope of the work, let alone the value it creates. Not so McKinsey. To come up with its recommendations, the Firm bundles various services—from research to scenario development to planning—into a final authoritative strategy document. Whereas other professions are increasingly forced to "un-bundle" because their clients demand more transparency, McKinsey has successfully rejected insight into its "black box." Too much transparency would simply challenge the value proposition.

Consequently, the Firm notoriously draws a hard line between insiders and outsiders; it maintains a culture of opacity and swears clients to secrecy. The disclosure of any deliverables is taboo. The result is usually zero publicity, and the negative headlines only occur when bad advice gets exposed. From Time Warner's ill-fated merger with AOL to General Motors' failed strategy against Japanese competition[16] to AT&T (McKinsey famously advised the company to consider cell phones as a niche market), McKinsey's publicized mistakes are well chronicled (most recently in the book *The Firm: The Story of McKinsey and Its Secret Influence on American Business* by Duff McDonald[17]), but they don't seem to have seriously hurt the firm's reputation—and certainly not its aura.

What is more threatening to McKinsey is the growing risk of commodification. Already, some consulting firms have begun to "templatize" some of their knowledge and "productize" some of their intellectual property. However, if McKinsey compromises its ability to provide more than just cookie-cutter solutions, it undercuts the very source of its unique value proposition. The minute a company has only one answer to any given question, it will soon find itself commoditized or outsourced. It is a sweet twist of irony that McKinsey, while preaching strategic clarity, analytical rigor, and efficiency, ultimately thrives on the elusive power of intangibles. In one of the most cutthroat and numbers-driven industries there is, ambiguity and secrecy remain priceless. Only the romantics survive.

Management science is reluctant to admit that businesses need ambiguity, but romantics know that it can be an essential factor in long-term success. The firm of fashion designer Eileen Fisher is a prime example. Her work is tailored toward women of all ages who prefer effortless clothing and a sense of elegance without drawing too much attention to their style choices. The culture of her organization is perfectly in keeping with this design philosophy, known for its understatement: a toned-down style that celebrates distinction over differentiation. In a *New Yorker* profile, Fisher described the units of her company structure: the "core concept team" and "the leadership forum," as well as a "different kind of leadership" where the leader "facilitates the process" and "holds the space."[18] The article notes that Fisher's description reads like a hermetic text: "Whenever the workings of the company came under discussion, the language became peculiar and contorted, as if something were being hidden." Even Fisher's VP of communications, Hilary Old, who participated in the conversation, was "equally powerless to explain the inexplicable."

Business Romantics recognize that ambiguity and secrecy are effective ways to foster a sense of identity within an organization. Or even transform it. Instead of inclusive "get on board" workshops or crowdsourcing to harness collective brainpower, try doing something more exclusive, more mysterious, like what we did at NBBJ, a design and architecture firm.

We identified thirty "thought leaders" across all of NBBJ's offices and invited them to "join nbbX." We positioned it as the "thought leadership platform of NBBJ," but in reality nbbX was something else: a badge of honor for the most prolific thinkers in the firm, an exclusive, handpicked community with rituals and symbols, a special club with strict rules of engagement and a high degree of interpersonal connections and mutual trust. In other words: nbbX was a secret society.

To launch nbbX we organized a one-day "lab." We had only one design principle for the day: make it feel like a wedding. We indicated to our colleagues that it was an honor to participate in nbbX; we expected them to commit to the full day, from nine in the morning to nine in the evening. Only adding to the suspense, we never made the purpose of the meeting explicit, allowing our guests to fill in the blanks with their own imagination: Would this be just another whiteboarding exercise (but with better food) with report-outs from breakout groups? Or would we be joined by surprise guests—magicians, monks, gurus—for a collective meditation? Or was this a revolt or a spinoff? The anxiety and excitement were palpable. More and more colleagues asked to be invited (we turned them down politely), and the sense of anticipation grew stronger as we got closer to the day of the event.

Creating a secret society at NBBJ, which was renowned for its highly inclusive and democratic culture, was anathema. For decades, NBBJ had done well as a business-savvy architecture firm with a seventy-year-old legacy and high-profile clientele. It exhibited a strong *persona* based on a widely embraced set of values and behaviors, but it didn't have an *agenda* (I always look at these two aspects when I assess the inherent potential and strategic trajectory of an organization). NBBJ lacked a distinct argument that went beyond the architectural practice. It was stuck in the middle between tangible craft and loftier goals, industry and intellect. The firm was keenly aware of its responsibility—to improve human lives at scale through the built environment—and yet many people at the firm were quietly afraid to voice an opinion outside of the architectural domain. Humble and soft-spoken, the company had a long-held bias against talking about its work, and it had strong cultural reservations about entering arenas of contentious discourse.

This humility was noble, but also a handicap. Today's professional services firms, or any firm for that matter, are confronted with markets of unprecedented complexity: everything from the intricacy of stakeholder grids to the deeply interconnected sociopolitical issues of our time is becoming more complicated. The innovation consultant Adam Richardson speaks of "X-problems": hard to pinpoint, hard to articulate, and incredibly hard to solve.[19] A company like NBBJ can address X-problems in two ways: through its work or through its thinking—ideally both. As I argued earlier, writing about McKinsey, a service whose value is typically captured in billable hours runs the risk of becoming commoditized. NBBJ had no choice but to assert itself as a network of thinkers with ideas it could extend to its clients and beyond. The main opportunity for us was to identify the root causes and greatest levers for positive social change. The main challenge was to be a design firm and *more* than a design firm at the same time. It meant asking questions for which we did not have an answer (yet).

The rich dialogue that took place in the nbbX lab gave our thirty colleagues the implicit permission to be in over their heads and have an opinion on the topics that mattered to them: as architects, but more important, as human beings. Immediately following the lab, we launched a blog, began to host lunch conversations, brought in outside experts, and established other conversational formats: all of which helped to make NBBJ more comfortable with the idea of pushing itself outside of its comfort zone.

The *X* in nbbX stood for just that: uncharted territory, open-endedness, the unseen, the adventurous and uncertain. Moreover, it also marked the reach of our initiative into and out of the organization: from questions to opinions to points of view to ultimately an ambitious intellectual agenda. We "hacked" the venerable NBBJ brand—the four letters represent the initials of the founders—replacing the *J* in the firm's name with the *X*. We created a romantic space: an *X* for double—or more—meanings.

These types of secret encounters are equally provocative outside the confines of our own offices. Imagine you're struggling with the design for a new product feature, and a mysterious stranger at an unusual business event offers up advice. At the end of the session, you discover the stranger is, in fact, Jonathan Ive, Apple's famed design leader. Welcome to House of Genius: an event series for business leaders and entrepreneurs that uses anonymity to create better conditions for collaboration and ideas exchange.[20] A typical House of Genius event brings together fifteen to twenty participants and three speakers who each give a five-minute presentation on a key challenge or business problem. At the start of the session, participants are asked to refrain from disclosing their backgrounds ("first names only") so that ideas are accepted at face value, encouraging everyone to contribute equally, without ego or social bias. This level playing field—along with the refusal to disclose content in advance—enables dialogue built on gut reactions and removes any filters that might impede the creative process. At the end of the meeting, there is the great reveal of identities, often accompanied by the sound of many "ooohhs." Jonathan Ive was in the house. Or not.

LIMITED EDITIONS

Secret societies such as nbbX or House of Genius are important for businesses and their leaders, but they are also essential to consumer experiences. Secrecy creates space for our true desires (and not only our "needs"). More specifically, anonymity permits us to speak freely and feel fully without the fear of real-world consequences or threats to our reputation.

The Internet, in particular, has scaled the value proposition of anonymity: the more we show ourselves through ubiquitous social media, the more we look for places to hide. After message boards (4chan), question-and-answer networks (Ask.fm), and confession sites (PostSecret), inevitably, there are now "apps for it": Whisper, which launched in May 2012, enables users to anonymously share messages with people in proximity to their location. And at the moment of writing this book, Secret had just emerged as the latest must-have anonymous-sharing app. Surrounded by accusations of providing a forum for cyber-bullying,[21] Secret allows you to make confessions or post other statements to your existing social network contact lists—anonymously. At the bottom of each "secret," others in your network can see where it was posted, how many people have commented on it, and how many people have "hearted" it. The more hearts, the more the secret will spread across your network and then across your friends' networks. This principle reveals an interesting paradox: the stronger the social currency, either through social ties or content popularity, the more endangered the anonymity becomes. But then, danger is precisely the point, isn't it?

By putting us in close contact with risk, exposure, and infamy, Secret satisfies both our longing for anonymity and our desire to be seen. It exposes our ambivalence with regard to privacy and publicity: we romantics like to be anonymous, yes, but only if others are watching.

Secret also illustrates another hallmark of the post knowledge economy: the experiential value of not knowing. By design, the app trades in the currency of doubt, enabling all kinds of trickery and "guessing games." The user's blind spots are its greatest asset. The things shared on Secret (e.g., "I work at Evernote, and we're about to get acquired," or "I slept with my best friend's wife") may be rumors or actual confessions; they may not be true, but they certainly pique our curiosity, especially if they come, anonymously, from someone in our network. We are suckers for what we don't know. Knowledge is power, but *not* knowing might be the more powerful experience.

This is also where the Secret Cinema series, founded in the UK, excels. Secret Cinema has completely reinvigorated the traditional moviegoing experience by making it social, immersive, interactive, and, most important, secret. Think of it as a blind date with a movie. It is a gathering for people who come together for "mystery screenings" of seminal movies, from *Casablanca* to *Alien* to *Blade Runner*, in unusual locations. Viewers don't know what they are going to see, although they are given some clues and subtle hints online before the event. What's more, they are warned *not* to spread the word! Not everyone complies with this request, of course, which provides Secret Cinema with a massive sales force.

The buildup to the event only heightens the excitement and eagerness to participate. The shows themselves often take on a life of their own, incorporating live performances, music, and food. At a screening of Wim Wenders's *Wings of Desire*, for example, a trained trapeze artist performed in sync with the trapeze scene in the film. For *The Shawshank Redemption*, Secret Cinema converted the Cardinal Pole School in the East End of London into a prison. For the showing of Ridley Scott's *Prometheus*, Secret Cinema turned a warehouse into a spaceship, and participants into boiler-suited crew members. Fourteen thousand people, over the course of several weeks, paid £35 for a film that turned out to be *The Battle of Algiers*, which most of them had probably already seen.

Secret Cinema morphs alternate reality games with flash mobs, if you will. Recently, it has begun to stage "immersive screenings" of new releases, showing, for example, Wes Anderson's *Grand Budapest Hotel* even before its theatrical release.[22] Founder and CEO Fabien Riggall believes "Secret Cinema is the future of cinema." We may not go to *see* movies anymore; we may go to *experience* them.

This may describe the future of music as well. In 2012, pop star Beck released his album *Song Reader* as "sheets of music" as a beautifully packaged physical product. Instead of recording any of the twenty new songs himself, Beck invited his fans to record them and then share them online. By doing so, he created artificial scarcity: he himself would only perform the songs in concert, while the lack of a digital product created a new and obscure market for "cover versions." This format allowed fans and musicians all over the world to (co-)create their Beck album while making Beck's concerts even more attractive.

Romantics savor the ephemeral, the magic of the fleeting moment. Nothing chafes us more than a poor-quality video of a memorable live event, a pathetic simulacrum of the iconic "you had to be there" moment. This is why live events such as Pop-Up Magazine, a curated series that bills itself as the "world's first live magazine, created for a stage, a screen, and a live audience," doesn't have any archived performances. Indeed, the organizers promise: "Nothing will arrive in your mailbox. Nothing will go online. Nothing will be filmed or recorded. An issue exists for one night, in one place."

In an era where everything we say and do is recorded permanently on the Internet, ephemeral media are on the rise. Just a few years ago, the picture-sharing app Snapchat, which self-erases pictures after a few seconds, would have been unimaginable; now it is the gold standard of the new wave of ephemerality.

A group of playwrights in New York City even used ephemerality as the design principle for their entire enterprise. The group, called 13P for the thirteen playwrights involved, titled their organizational structure "The Implosion Model." Each of the thirteen playwrights was given the opportunity to oversee their own production. When all the playwrights had served at the helm of one production—thirteen plays in all—the company vowed to disband. By explicitly stating its shelf life, the group made clear that its focus was on the work, not the institution.

"The long-term goal was not to make a profit. It was simply to sacrifice to make something. Then, when those things—the shows—were made, it was over," Rob Handel, one of the founders, told me.

In an interview with the *Huffington Post*, another member of 13P, Young Jean Lee, stated, "It was one of the purest artistic experiences I've ever had. I have my own organization and the fact [is] that I have to worry about the sustainability of continuing to pay my employees and keeping a business going. I don't have this pure artistic experience where I can do whatever I want. It just doesn't work that way."[23]

"What can you do when sustainability is not your goal?" Madeleine George, cofounder, added. "It opens up all sorts of exciting opportunities."

In 2012, after almost ten years of award-winning productions, members of 13P gathered together at a farewell fête at the Public Theater in New York City. Implosion was the theme of each of the celebratory toasts, performances, and speeches. And then the countdown—5-4-3-2-1—13P was, in an instant, no more.

NEGATIVE SPACE

My daughter and I recently visited a toy store in my neighborhood because the owners had arranged for a special treat: a session with Karl Johnson, one of America's foremost silhouette artists, who would create cutout portraits. Demand was high, and there was a long wait. Before my daughter finally took a seat for her slot as "model," I used the time to observe the artist's work.

Johnson sat in the store on a small chair from nine A.M. to six P.M. that day without any breaks, creating close to sixty portraits. He asked the children to sit still, offering up some friendly small talk to make them feel comfortable. The entire time he was talking, he was also freehand cutting. His scissors moved relentlessly on the paper, but his eyes never left his model. He scanned for the contours of the child in front of him and then translated that visual pattern into scissor cuts. I was struck by the complexity of his task and, at the same time, by its simplicity. "Silhouette portraits capture the subject's essence—who they are—in the most simplistic of ways," Johnson wrote in a post.[24]

Later, I read that he attributed his unique talent to his innate monocular vision. He sees with only one eye, his right. He says this forces him to assess the distance and shape of an object by examining its shadow, meaning that he sees the entire world as a series of silhouettes. In this way, he carves out meaning *ex negativo*, by excluding, or cutting out, other information. He creates the perception of three-dimensional depth through some seemingly random lines, able to reveal the uniqueness of one person—the human spirit—in the blink of an eye.

Watching a master like Johnson at work, I was reminded of the way that certain brands have cultivated an artistry of negative space, carving out just enough visibility to render themselves absent. Like an afterimage, they remain visible in our minds through their very departure, their emptiness, and their restraint. Only their silhouette remains. How does a brand artfully disappear?

Cult fashion brand Maison Martin Margiela (MMM) can offer some clues. The house has remained swathed in anonymity throughout its twenty-year history, and its namesake designer, Martin Margiela, is often represented only by an empty seat at his own fashion shows. The fashion world has grown accustomed to such behavior from the elusive designer. Where others cater to the demand for ubiquity and larger-than-life personalities, he trades on his invisibility, making it an unexpected asset. Margiela, a graduate of Belgium's Royal Academy of Fine Arts, spent the early eighties as an assistant to Jean Paul Gaultier. At the time, fashion in Belgium, and Antwerp in particular, was deeply inspired by the more radical of the deconstructionist designers such as Rei Kawakubo of Comme des Garçons. By joining forces with a group of like-minded designers—referring to themselves as the Antwerp 6—Margiela started to bring the subversive ideas of deconstruction to Europe, ripping apart clothes and exposing seams. Today, so many of the styles that we take for granted—shredded jeans and the use of recycled industrial materials in clothing, for example—can be directly attributed to his influence. Instead of crafting a story around its clothing line, the characterless facade became the essence of the brand; in fact, runway shows often had "faceless" models with large wigs covering their entire heads.

To this day, this cult of *im*personality spreads through the aesthetic of the brand: stores are never identified with signage; staff at stores and at MMM headquarters wear standard white lab coats; white—called "whites" in MMM-speak—is the ubiquitous color of all stores and of the sheets that cover all in-store furniture and displays; packaging is monochrome and logo-free; and seating is mostly first come, first served, avoiding the industry standard of seating hierarchy. It has been impossible even for industry insiders to reach Martin Margiela directly. All company requests must be done by fax addressed to "Maison Martin Margiela." Everything about MMM's creative process communicates the potency of absence. As a result, the acolytes and avid fans of the brand take on the role of spokespeople. They share the "inside scoop" that even the brand itself refuses to acknowledge. Absence serves as implicit invitation to participate.

In 2009, MMM majority stakeholder Renzo Rosso publicly admitted that Margiela had not been seen at the offices for a long time: "He is here but not here."[25] A few months later, a press release announced that Margiela had left the business, but that no creative director would be appointed to replace him.[26]

Similar to MMM's design principle of absence, the Japanese retail company Muji, heralds the virtues of emptiness. As the son of a Shinto priest, Kenya Hara, Muji's art director, places this emptiness within a rich context of Japanese Shinto religion, specifically the Shinto shrine. He told an audience at a conference:

> There is no way to make an appointment with the Gods. The only thing we can do is to invite the Gods as guests. There is a curious type of structure in Japan that is common to us. It is called Shiro. It is made of four pillars arranged in a square with their tops joined by a straw rope. Inside there is nothing. That is, it is empty; in fact, it is emptiness. If we create a condition of emptiness, the Gods that are the forces of nature might come to fill it. Because emptiness is itself the possibility of being filled, the Gods which see everything couldn't fail to notice the empty space. But that doesn't give us any certainty that they will enter. They may enter. This may carry great weight. What people pray for is the possibility.[27]

"Emptiness is not a message," Hara concluded. "To offer an empty vessel is to pose a single question and to be forever ready to accept a huge variety of answers."

This reminded me of what the photographer Platon, famous for his portraits of world leaders, once told me about his work: "I try to be as empty as possible when I approach my subject. It's like I see them for the first time, and my empty mind allows me to truly recognize them as who they are."

Brands such as MMM and Muji, and artists like Platon, are inviting us to contemplate a business world where brand books and elevator pitches are beside the point. Their aesthetics share a concern with creating conditions, almost as if they are chemists putting the right elements together in a test tube or another receptacle before heating them with the Bunsen burner. They evoke a state of receptive—not passive—waiting.

This receptive waiting might well be defined—in romantic terminology—as waiting for the muse to strike. The Italian notebook maker Moleskine, which celebrated its IPO at a valuation of more than $600 million—a stunning anachronism in a world that usually glorifies tech start-ups and digital products—has created an entire business model predicated on the romantic notion of the modern muse. The legendary paper notebook rose to fame through artists such as van Gogh, Hemingway, and Picasso, but it had all but disappeared in 1997 when the small Milanese publisher Modo & Modo established the brand and business. The company began to market the notebook very cleverly as a tangible reservoir of the artful and playful, as the nostalgic diary for the ideas, insights, and emotions of the connected modern human.

The story behind this renaissance is romantic in and of itself: Maria Sebregondi, Moleskine's VP of brand equity and communications, told me she and two friends of hers had an idea during a sailing trip on the coast of Tunisia in the mid-nineties. Their vision was to translate the artist's old-fashioned sketch book to the age of digital nomads, appealing to today's knowledge worker on the go. They wanted to create a nostalgic physical artifact that was complementary to his or her digital devices. She admits that she simply followed her own passion rather than vying for, let alone carefully evaluating, the commercial aspects.

The Moleskine notebooks function as what Sebregondi calls "analogue clouds" and she believes their goal is to serve as an open platform for "shared intimacy and imagination." Similar to Muji, Moleskine's brand association and value proposition is emptiness. Emptiness is a luxury in a time that privileges connectivity and transparency over almost everything else. And yet Moleskine is not categorized as a luxury brand. "Unlike luxury, which is exclusive, culture is inclusive," said Sebregondi, "and Moleskine is a part of a popular culture that everybody can access." Moleskine notebooks are democratic vessels for self-expression, as well as handmade objects that serve as concrete, simple containers of our complex lives. The books provide the modern nomad with an emotional anchor, a home away from home, while at the same time serving as a device of curiosity, discovery, and exploration. They are the narrative compass for today's business traveler: rich with context (every book comes with an insert evoking Moleskine's history) and yet completely open for something new. The notebook's blank pages feel like an opportunity to reimagine life. Every new book is an unwritten story, and it holds the romantic promise of starting all over again.

Moleskine recently entered a partnership with organizing software provider Evernote to create a collection of "smart notebooks," and the two firms are in the process of launching other digital initiatives. I asked Sebregondi if this transition to the online world could jeopardize the purity of the experience and the integrity of the emptiness that Moleskine stands for. She is not worried: "The digital world offers an infinite amount of empty space."

The negative space that brands such as MMM, Muji, or Moleskine trade in serves other brands as well, especially those with a strong persona and the confidence to leave room for the unknown: the website of Creative Artists Agency (CAA), the legendary Hollywood talent agency, shows only the firm's contact info and nothing else. As a visitor, you get the message: CAA is the top firm in town, as exclusive as it gets, and confidentiality and class are its best and most important credentials. The site is a "veil" that only those in the know can uncover. Like a secret society, the shell is flat and unimpressive, making the hidden core all the more mysterious and attractive.

And then there is the Evian brand: using aura alone, it turned the world's most ubiquitous commodity, water, into a luxury item. The product is pure fantasy: the clear surface of Lake Geneva; a glass at a dinner table in Paris; a cool bottle during a marathon. Evian is the epitome of the *je ne sais quoi.* Like all great romantic brands, it is an invitation to imagine. Such is the romantic's credo: the proof is in the promise, not the pudding.

Transparency and openness have become buzzwords in recent years. Open government; open source; open innovation; open leadership—the list goes on. From the ecological footprint of products and information about the supply chain, to employee feedback, to product innovation and firmwide strategy, companies have begun to outshare one another in order to gain the trust of their stakeholders.[28] It is a commonly held belief that all this openness increases legitimacy, accountability, and access to ideas, resulting in improved performance. But it is important to realize that it comes at a price. The art critic and curator Deyan Sudjic writes:

> The symbol of transparency is a two-edged sword. In transparency we manifest democracy and clarity—shining light on dark spaces. Yet when you look at a transparent glass wall in daylight, what you see is a reflection of yourself. If we are to live in a world where we continue to encourage innovation, we need the messy vitality of opacity.[29]

I love that phrase, "messy vitality of opacity," which to me is the hallmark of great creativity. After all, closing a proverbial door—whether it is the door to your office or the door to a brand—is not always a bad thing. In their more innocent forms, secrets engage our empathy, our curiosity, and force us to consider what we hold most dear. Excessive use of transparency not only threatens our concept of privacy but also the romance of the hidden.

As a response to radical transparency, Business Romantics use mystique to create romantic spaces that allow us to enjoy what we can't see: pop-up shops that appear unannounced; restaurants with secret locations that can only be found by word of mouth; or "dining-in-the-dark." All these new services purposely limit one dimension of the user experience, like cutting off one leg of a stool. Snapchat, the picture-sharing app, obscures memory by denying online permanence; Secret obscures the identity of the user; the Pop-Up Magazine obscures all-time availability; and Secret Cinema, the mystery screening series, obscures the content by not revealing in advance the movie it will show. By design, these opaque and ephemeral formats represent a new breed of media and services for the post-knowledge economy. It is an economy in which the overwhelming abundance of facts and figures available online—clarification—is only increasing our desire for obfuscation. Welcome to the "misinformation economy."

Our romantic business models, brands, customer experiences, and charismatic leaders thrive on the unknown, on the luxury of not having to share everything. Business Romantics prefer understatement, silence, emptiness, and even absence to dogmatic transparency and always-on presence. We refuse to explain or reveal certain things—not because they are sacred, but because we *want* them to be sacred. Things that don't last, last longer. Experiences in the dark illuminate our desires. If you want to highlight something, hide it! Invite secret agents, undercover projects, and mystery into your organization. Try, for once, to shut some doors. The things we lock away, the stories we keep for ourselves, the shadows we cast, the spaces we cut out: they make us want more.

CHAPTER 9

Break Up

> I'm not sentimental—I'm as romantic as you are. The idea, you know, is that the sentimental person thinks things will last—the romantic person has a desperate confidence that they won't.
>
> —F. SCOTT FITZGERALD

The performance artists Marina Abramović and Ulay began an intense romantic relationship in the 1970s, living and working together in a van. When they sensed the relationship had run its course, they decided to walk the Great Wall of China, each from opposite ends, with the goal of meeting at the halfway point for one last kiss. Abramović started from Shan Hai Guan, on the shores of the Yellow Sea, and Ulay from Jai Yu Guan in the Gobi Desert. After walking 1,500 miles, they met in the middle and kissed, then turned their backs on each other and did not see each other again for almost thirty years. This well-choreographed final kiss marked the end of their journey together. It was their full stop.[1]

Great symphonies never peter out slowly—*plink, plink, plink*ing—while the audience begins wandering away. There is the grand flourish, the triumphant final wave of the conductor's hand, that tiny but seemingly infinite moment of silence, and then the final space and invitation for applause. The only way to end anything is to deny an audience the opportunity for everything. A powerful ending requires the courage and conviction to say, "I gave you all of myself. There is no more. Good-bye." Such endings—the full stop—do not mean that there are no open questions; they do mean, however, that the curtain is closed.

When it comes to business, we tend to underestimate the power of this endnote. We understand beginnings—they might intimidate and challenge us, but they always promise the thrill of novelty: yes, we are alive! Yet we are often confused and even miserable after endings. Whether in our leadership, employee, or customer experiences: ending well can be difficult. With so little practice, we simply lack the required skill set, and we fail to anticipate the emotional toll of parting ways. A business divorce is still a divorce, including hurt feelings and all of the associated collateral damage. A postmortem after an intense tenure or project is so much more than a series of dry questions, so much more than a checklist of a "went well/did not go so well" analysis: it is a moment for grief; in fact, the completion of a tenure or project can often lead to varying levels of postpartum depression.

How do we honor our moments of closure in our employee and customer experiences? And those in our own careers? We need more designs for (un)happy endings, more rituals for what Susan Sontag described as "perpetual acts of separation and return."[2]

Romantic leaders can find creative ways to consecrate periods of transition. In 1995, the chairman of Samsung, Lee Kun-hee, was trying to transform the low-cost commodity producer of electronics into an innovation leader in the mobile phone market. Instead of distributing memos or calling together management to discuss the brand and company values, he directed Samsung's factory workers to build a mountain of 150,000 flawed mobile phones in a field outside the company plant. Thousands of staff members were instructed to stand around the pile and bear witness as the entire mound of technological innards was set on fire. When the flames finally fizzled out, bulldozers were brought in to clear the detritus away. The message was clear: only excellence in mobile phone technology would reach the marketplace. Anything less would go up in flames.

Today, Samsung is the world leader in smartphone sales. In the fourth quarter of 2013, its Galaxy phones outsold even Apple's storied iPhone.[3] "The Great Phone Incineration of 1995 functions as a legend within the company. It's a major point in the history of Samsung and of the chairman," said Sam Grobart, a reporter for *Bloomberg Businessweek* who recently profiled the notoriously elusive company. To this day, new employees at Samsung are told about the event, and it serves as both a point of inspiration and of fear. "The power of myth is regularly deployed around Samsung," Grobart observed.

Samsung's "endnote" is reminiscent of the legendary mass burial of a product flop by game producer Atari.[4] In 1982, the then-fast-growing company released its video game *E.T. the Extra-Terrestrial*. On the heels of the blockbuster movie, it was supposed to be a slam-dunk hit. Except, it wasn't. The initial sales were respectable, but soon users complained about the poor experience, and finally *E.T.* was taken off the market. Until today, it is considered "the worst video game ever," and the episode contributed to sending the global video-game industry into a historic tailspin, known as "Atari shock" in Japan.

And what did Atari do? Along with unsold copies of other product flops it suffered that year, the company bought back millions of cartridges, shipped truckloads into New Mexico under cover of darkness, and buried them—encased, as if radioactive, in concrete—in the desert, near the detonation site of the first atomic bomb. Like Samsung's Great Phone Incineration, Atari's mass burial combined the practical with the symbolic: the unorthodox action is believed to have allowed Atari to write off the disposed-of material for tax relief purposes. But for the romantic, the most remarkable aspect of this operation is its ritualistic expression of corporate shame—the attempt to hide a high-profile failure with a secret funeral.

Moments of closure and transition occur every day in customer interactions as well, although few of them involve incinerations or undercover burials. Consider the airline industry: many carriers are struggling to get the beginning of the experience right—delays, long lines, and last-minute gate changes are only a few of the inconveniences—but many of them master the power of the endnote. The flight attendants will not only announce the end of the trip and send passengers off with a routine "Thank you for flying with us. Have a great day or onward journey," they also line up at the exit door to personally say good-bye when the plane reaches the gate. It is customary for the pilot to make an appearance and bid farewell, which is another nice gesture. There is a sense of relief for everyone involved: "We arrived safely. It meant a lot to us that you gave us your trust, and we enjoyed serving you—good-bye." Dutch airline KLM even hands out very special parting gifts to business-class passengers: miniature ceramic houses that reveal themselves to be containers of schnapps.

This moment of farewell to (hopefully) returning customers is a key moment in all customer experiences, not only because last impressions frame the perception of service quality as much or even more so than first impressions, but also because the business is essentially concluded at this point. Any attention after the actual transaction is a gratefully received bonus. How can you craft an "exit strategy" for your business—the point in time or place where the transaction ends—to make sure that you send your customers off with full hearts?

Arguably, it is easier to do that when you know your customers will return. But if they are leaving you for good, then an excessive use of kindness is even more precious. Instead of trying to win your customers back—inquiring about the reasons for their departure or aggressively courting them—try accepting the end of the relationship and celebrating it in style: honor your "ex" with an expression of acknowledgment and gratitude.

Carefully designing parting moments is equally important at the end of employment, whether it is a firing or a resignation. Business school doesn't prepare us for the intensity of emotion that occurs during these experiences.

Rowan Gormley, founder and CEO of Naked Wines, left his job at Virgin in 2010. He had been at the company for more than ten years, working closely with Virgin founder and CEO Sir Richard Branson to incubate Virgin Wines. After the unit was sold to a third party with whom Gormley didn't see eye-to-eye, he was fired. In a fierce act of rebellion and solidarity, Gormley's twelve employees simultaneously tendered their resignations by e-mail, then went down with him to the nearest pub and decided to launch their own business together. They were at once excited and frightened about what would come next, but they were comforted by the fact that they were still together and had kept their values and principles intact. Gormley described this to me as one of the most meaningful moments of his life: "With so much to lose, so much to gain, everything was possible." Naked Wines, the new company he founded with his former Virgin colleagues, became profitable in the U.K. after four years, and in 2012 Gormley moved to Napa Valley in California to launch the business in the United States.

Gormley chose to leave with a gesture of bravura showmanship; others prefer a quieter endnote exemplifying their honor and respect for the organization. Once again, sports can provide us with a blueprint. Frank Rijkaard, a former coach for Spanish soccer club FC Barcelona—or Barça, as the fans call it—is a perfect example. At the beginning of his five-year tenure, he won two national league titles as well as the prestigious Champions League, but he failed to get any silverware for the proud club in his last two years. In May 2008, a humiliating 1–4 defeat against archrivals Real Madrid marked the absolute low point of a tempestuous season. This left Barça's then-president Joan Laporta no other choice than to dismiss him. Everyone knew that Rijkaard's time in Barcelona was over. Yet Rijkaard managed to maintain his optimism and integrity in the face of the unfolding tragedy. He continuously defended his players and never criticized them in public. Instead, he absorbed the anger and frustration of the fans with an unshakable work ethic that was always directed toward the next game.

In his last game for Barça, a standard victory over a second-tier team, young striker Giovanni Dos Santos scored a hat trick (three goals in one game). Each time the ball hit the net, each goal eclipsing the previous one in deftness and beauty, the teenage player turned toward Rijkaard. The coach, normally known for his stoic comportment during games, was on his feet, all smiles, lifting both hands in the air and cheering as if his team had just won the Champions League final. It was a remarkable moment, a moving outburst of joy in the twilight of his tenure. Rijkaard understood all too well that, in the end, dignity mattered more than trophies—especially for a club that prides itself on being *més que un club*, or "more than a club," as the motto goes. In his last press conference Rijkaard simply said: "It has been an honor to work for this club."

We need more of these honorable exits and closures, transitions that allow us to express our best selves. Take the departure of

Groupon's founder and CEO, Andrew Mason, in 2013. When he left the helm of the deal-of-the-day website, Mason wrote a letter to the "People of Groupon" that began with: "After four and a half intense and wonderful years as CEO of Groupon, I've decided that I'd like to spend more time with my family. Just kidding—I was fired today."[5] While the opening of the letter felt more like a practical joke, Mason's main point was that he was justly fired. He explained to his more than eleven thousand employees that he had lost sight of his intuition for the company's growth and that a new leader would give them "breathing room." It was a remarkably graceful exit—both self-effacing and generous—because it encouraged Groupon to move forward with confidence and a clear vision. He asked his former colleagues to embrace the new CEO while exhorting them to keep striving for excellence.

Mason's letter stands out in corporate culture. More often than not, departures involve banal euphemisms, mixed emotions, finger-pointing, and considerable awkwardness and fear from all parties involved. The benefit of being fired is that the end is definitive, and in this case the ruthlessness of corporate protocol may come as a blessing. The full stop here is palpable, made visible to everyone by humiliating and disempowering rituals: cleaning out the cubicle and hauling away the contents in a cardboard box, for example. And yet the minute the fired employee exits through the revolving doors, a process of genuine reflection can begin. At least for the person leaving. The firing manager, on the other hand, is expected to move on quickly because business logic dictates so. After he had let go of a dozen staff members, I remember one of my former bosses repeating "I have a business to run," as if he were seeking absolution.

Resignations are also trickier than you might think. Even when we leave on friendly terms, a graceful exit that feels good for everyone remains the exception to the rule. We are afraid of losing control over our narrative. There is also the daunting task of untangling the complex web of relationships and the institutional knowledge of the company we have built over time.

In my career, I resigned three times from posts that I held, and each time I underestimated the emotional aspects of the separation. My resignation from Aricent was the most extreme experience. In 2010, during my tenure as chief marketing officer of Frog Design, I was tapped to also head up marketing for this IT outsourcing company, Frog Design's parent. The Aricent investors, a private equity firm, were making a strategic push to integrate the engineering capabilities of Aricent—an organization of ten thousand employees based almost entirely in India—with Frog's cachet in creativity and innovation. The management team and board thought my understanding of branding as well as my familiarity with the quirky Frog culture would enable me to move the entire organization upmarket by harmonizing the brand architecture.

Even on paper, this sounded like a Mission Impossible, but the challenge was too good to pass up. In my new position, I would need to find common ground between two radically different organizations: the engineering culture of Aricent versus the design culture of Frog; pocket protectors versus hipsters; reliability versus creativity; scale versus intimacy. My job, in essence, was to transcend both cultures: to carve out "the third way"—a third meaning, one that would serve as foundation for a unified story in the marketplace.

As Aricent's chief marketing officer, I traveled to our locations in Gurgaon, Bangalore, and Chennai several times a year. On each visit, I would sit in office park courtyards at different business triangles or shuttle between office and airport, becoming intimately acquainted with my driver as well as the smell of the vanilla tree air freshener in his car. The entire effort reminded me of the Olympic Torch Relay: I was visiting, time and time again, to shake just one more hand, to get the ear of one more employee at a local town hall. After a year of campaigning, we launched our new brand—the "Aricent Group"—with great fanfare in India and our other twenty locations worldwide. I started receiving photographs from people all over the globe. In every single one, employees were holding up orange Aricent balloons and cheering. We had made the impossible possible.

The enthusiasm, however, was short-lived and eventually gave way to conflicting agendas as the engineering and design mind-sets began to clash. The company struggled with sustaining collaboration among thousands of employees halfway across the world. Eventually sales faltered, and institutional support wavered. The board reversed course, and our mission was dead.

On my last trip to India, I remember sitting in the lounge of the Delhi airport at three A.M. with my colleague and partner in crime, the head of sales, who like me had been tasked with integrating the two organizations. "To the end of a dream," he said as we clinked our bottles of Kingfisher beer. I had come as a friend and left as a tourist.

I had failed—at least that is how I felt—and I didn't have a reason to stick around. I decided to resign from my position and called my boss, the Aricent CEO, to tell him that I would like to go back to my former role at Frog Design. With the wind out of my sails I couldn't wait to jump ship. But I also felt the responsibility to help with the transition. Over the course of three months, I slowly detached myself. Instead of a symbolic endnote, a kiss at the Great Wall or a great incineration, my organizational farewell was prolonged and cumbersome. I died a death by a thousand good-byes, more painful than any of the many pain points I experienced during my conflict-laden tenure. I phased out, my authority faded, and my investments and connections weakened before I finally disappeared. As my own story illustrates, it is important to understand that romantic endings depend on "when you stop your story," to paraphrase director Orson Welles.

When it comes to designing our breakups, we can take inspiration from the weaving techniques of the Navajos, a native American tribe. All Navajo textiles contain a thin, horizontal line that usually extends from the center design field to the outer border, contrasting with the outer area but often of the same color as the center field. In the Navajo language, the term for this weaver's pathway is *ch'ihónít'i* or *'atiin*, which translates as "way out" or "road" respectively. This pathway, also called "spirit line," is meant to release the energy woven into the textile in order for the weavers to retain the energy needed for weaving other pieces.[6] The break in the design allows the weavers to emotionally separate themselves from the woven product so it can be sold. The deliberate flaw prevents them from full mastery and keeps their imagination and inspiration alive. The design is broken so that the spirit can stay intact. The weavers, who have given everything and put it into their textile, are given back. Their product is never perfect, so they can begin again.

In our jobs, we may have plans for the first one hundred days, but rarely for the last. We often forget to design for the "way out." And then things break before we can break (them) up.

The most artful endings make genuine space for reflection and mourning as well as for the promise of renewal. They can summon forth our most mythic selves—like Samsung's Great Phone Incineration—or they can remind us of our better natures—like Andrew Mason's departure from Groupon and Frank Rijkaard's final moments with FC Barcelona. Romantics know to honor each and every one of these endnotes.

Clink two Kingfishers together and tell the world that the dream is done. Weave a "spirit line" into your customer experiences, projects, and relationships, and into your career. Hone your muscles of separation and stay sharp: at least once a year, break up with a colleague, customer, partner, or brand. Appreciate break-ups as preemptive strikes against disillusion and disenchantment. Quit! Leaving makes us confront what we really want: it is the ultimate act of romantic defiance. As romantics, we either give everything—or nothing.

CHAPTER 10

Sail the Ocean

If I get married, I want to be very married.

—AUDREY HEPBURN

I recently had lunch with a friend who is an investment consultant with a large health care company. When I asked her about her new position at the firm, which she started about a year ago, she could only look at me, sigh, and say, "The thrill is gone, Tim." We talked about what that meant for her in her daily life. "I live in airports; I'm always on the go, responding to clients at the drop of a hat. At the same time, the routines are becoming mind numbing: the TSA check-ins, the hotel minibars, the meetings in all the same restaurants in all the same cities. This job is making me feel old . . ."

My friend, a mother of three, fantasized about a work experience that would leave her feeling inspired, or at least energized. When she first started her work as a consultant, she enjoyed connecting with people and helping clients achieve their goals. There was a sense of excitement, glamour even, dashing off at a moment's notice to different cities around the world. Her suitcase, packed and prepped in her bedroom, was always at the ready.

"When I was younger, I was very inspired by [the writer] Joan Didion's packing list in *The White Album*,"[1] she admitted to me. "It left a really strong impression on me." Didion's list, including items for dangerous reporting trips, was written with cool precision and her signature irony. My friend kept a copy of it in her bedroom closet for years when she first started her work in consulting.

To Pack and Wear:

2 skirts
2 jerseys or leotards
1 pullover sweater
2 pair shoes
stockings
bra
nightgown, robe, slippers
cigarettes
bourbon
bag with: shampoo, toothbrush and paste, Basis soap, razor, deodorant, aspirin, prescriptions, Tampax, face cream, powder, baby oil

To Carry:

mohair throw
typewriter
2 legal pads and pens
files
house key

"I know it sounds ridiculous because my travel and my work had nothing whatsoever to do with journalism or Joan Didion, but somehow, just by associating myself with such an influential person, my work trips used to feel sort of exciting. Just by thinking of her packing list—I mean, bourbon, for God's sake! Cigarettes! Imagine getting that through TSA today!—I could associate myself with some kind of bigger narrative, some kind of exciting thread in the culture. All this, too, because I'm a working mom, and Didion has always been an inspiration to me as a feminist icon."

I told my friend that I thought she was a Business Romantic—albeit a closeted one—and she started to laugh. "Maybe so. But don't tell my clients that!"

Isn't it remarkable how a seemingly meaningless artifact like a packing list—when attached to the right source of inspiration—has the potential to lift our spirits? It can imbue the mundane experience of business travel with larger meaning, tap into the zeitgeist, and allow us to share our experiences with the world at large. For my friend, this meant a wink and a nod of admiration to Didion every time she was hassled by TSA for her tiny plastic vial of face cream.

And yet, if it only requires these small token gestures, why can it be so difficult to recapture the romance once it's gone? I asked my friend what it would take for her to feel that old sense of exhilaration in her current work situation.

"Oh, I don't know," she said. "I used to find someone like Joan Didion romantic because I was younger, and I was more adventurous. Now I have so many more responsibilities, and I feel like I'm pulled in so many more directions. When I travel, I just feel annoyed and kind of brain-dead. Work is about routines now; I feel like I'm sleeping my way through it."

What lies between the thrill of the first-time experience and our final moment of departure, our endnote? For so many of us who are somewhere in the middle of our journeys, it is a moment for reflection. Some of us become frustrated, cynical, or strident. Others may feel closer to a sense of mastery with work, to the deep satisfaction of performing "at the top of one's game." Whether we see this middle as a place for renewal and reawakening, or as a place from which to assess our growing sense of confidence and accomplishment in work, the middle offers romantics a great vantage point. We can look back at our earlier selves and wonder at some of our misguided and simplistic notions, or we can cast our minds forward into the future, planning for possible endings or transitions. It is a place where we can bemoan the loss of innocence or appreciate the opportunity to reframe the beginning and make more sense of it. Either way, the Business Romantic needs to make a choice: what are we committed to?

Age and experience play a role, of course. Social psychologist Heidi Grant Halvorson put it this way: "The older we get, the more we want to hang on to what we've already got—the things we've worked so hard to achieve. We also have more experience with pain and loss, having been knocked around a bit by life, and having learned a few lessons the hard way."[2]

Business Romantics navigate through new and often richer forms of satisfaction in the middle. The ecstasies of those early years are gone—the big promotion or the very first pitch or proposal—but they are often replaced by acceptance and a sense of ease with the rhythms and rituals of work. Perhaps most important, those of us in the middle know how to savor life's relatively few moments of relaxation. As Halvorson put it, "Happiness becomes less the high-energy, *totally psyched* experience of a teenager partying while his parents are out of town, and more the peaceful, relaxing experience of an overworked mom who's been dreaming of that hot bath all day."

It's no coincidence, for example, that most of our romantic heroes from that Swiss summer of love—Byron and Shelley, say—

died young. It's one thing to fly by the seat of your pants when you are in good health and full of potential; quite another when you need to think about health benefits, save for retirement, and set up college tuition funds. A recent study, published in the *Journal of Organizational Behavior*, points to differences in the way workers under thirty conceive of job satisfaction compared to older workers.[3] The results, garnered from a sample of more than three thousand technical professionals, showed that younger workers were more interested in opportunities and skill acquisition while older workers wanted more commitment from employees as well as a good work–life balance. Most interesting of all, however, was the fact that older workers were still keen on the excitement of job promotions and the acquisition of new technical skills. Yet they needed to have a clearer definition of their commitment to and from their employer.

Matthew Stinchcomb, the first employee after the founders at purpose-driven e-commerce site Etsy and now the company's VP of values and impact, shared with me his vision of workplace commitment: "I'm always looking for opportunities for employees to connect with our larger vision of reimagining commerce." Stinchcomb readily admitted, however, that some knowledge-worker jobs were more romantic than others: "If you're working in customer relations, answering one hundred e-mails a day from frustrated people, it's a lot harder to connect with Etsy's core values. That's why we also need to create a culture that gives employees other outlets for experiencing creativity and connection." He referred to one of the company's core projects—Etsy School—which offers employees time and space to share their gifts and passions with other colleagues at work. The school uses the company's resources and space, and many of the classes are taught during work hours. By folding outside passions into the daily rhythms of the workday, Etsy commits to opportunities for self-expression within the workplace.

Stinchcomb wrote the Etsy company values, and he made it a point to openly address the way our commitments ebb and flow throughout our lives at work. "There are certainly times in the day when I grow frustrated or bored," he told me. "I think that's just human nature. I see it for what it is, and I try to address my fear: 'what if someone realizes that I'm not engaged?' In those moments, I try to say, 'I'm going for a walk.' I don't want to mess around on the Internet, trying to look like I'm busy."

Stinchcomb crafted the final value for Etsy—"Keep it real, always"—to capture the company's commitment to an authentic dialogue with its employees. For romantics, "keeping it real" means admitting that sometimes we cannot conjure up the romance. At other moments, it means acknowledging our fears of incompetence or disengagement. Instead of hiding in denial behind busywork, we face down our fears and name them. And, as with everything in life, we accept that we will plateau from time to time. Even Business Romantics will spend periods of time in the middle, in breathers between the highs and lows. If we are "keeping it real"—if we are honest with ourselves about where we are at and where we are committed to going—these valleys between the peaks put our experiences into perspective.

An exploration of commitment inevitably brings up the conundrum of modern relationships. Nowhere is it better rendered than in Richard Linklater's trio of films exploring the evolving romance between a typically volatile pair of lovers: Jesse, played by Ethan Hawke, and Céline, played by Julie Delpy. In the first film—*Before Sunrise*—we watch the two fall in love during an intensely intimate night together in Vienna. They lose touch only to be romantically reunited a decade later in *Before Sunset*. The third film—*Before Midnight*, released in 2013—looks in on Jesse and Céline in "the middle." They are now well into their forties, with two children, a mortgage, and a veritable minefield of issues. All three collaborators on the film—Linklater, Hawke, and

Delpy—discussed their creative process in an interview in the *New York Times.*[4] "We had an idea to just show them on a typical day: someone goes and picks up the kids, and it's only later at night that they have this together moment," Hawke said. But the three artists dismissed that idea because it didn't truly evoke the romantic spirit of the previous two films. They wanted the third film to harken back to romantic possibilities while also staying true to a middle-aged relationship. As Hawke put it, "You can't keep having first love forever."

What can we learn from commitments that are no longer fresh and new but "long in the tooth"? Jesse and Céline can no longer project images upon each other, hoping that the other will fill out their deficiencies. It becomes increasingly difficult to tell a story about someone after you get to know them, after you're no longer strangers. Intimacy fragments the narrative.

Still, I make an argument for rekindling romance with a person, a company, or a career while in the middle. How, in our professional lives, can we sanctify this middle with commitment? Instead of simply walking out the door, how do we find a way to fall in love again? The amount of time that average workers stay with their employers is decreasing, and the idea of a lifelong career with one company is a vestige of the past.[5] In our work lives, we are serial daters: always on the market, always looking for the next best thing. Sure, employment is not a marriage, but then again, we spend more time at work than with our spouses or partners. What would it mean to stop seeking the revolving door, to find satisfaction where we are?

We can make our commitments stronger by demystifying the siren songs in the greater landscape of business. Some of us are entrepreneurs by nature, finding our romance in each new company we create. Entrepreneurship, however, has its own "middle," and not all of us have the vocational calling to withstand it. Seth Matlins, who helped launch and grow the marketing consulting arm for

Creative Artists Agency (CAA) in Hollywood and served as chief marketing officer for concert organizer Live Nation, transitioned out of his corporate marketing career to spend time with social entrepreneurship initiatives. As a true romantic, he imagined the iconic moments of starting his own business, the early mornings spent on perfecting the brand story, the late nights in the garage packing T-shirts into boxes: no longer a CMO but an entrepreneur.

"I had this vision: 'I'll be like Jobs and Woz in their garage,'" he told me. "But that romance can fade really quickly. There are only so many shirts you can prepare for shipping happily and with real mindfulness. I found my tolerance for frustration was markedly lower in my own enterprise than it was when I was in the service business. It was so much harder." Ultimately, Matlins found his bearings in his new work identity, but he made it clear that the path of the entrepreneur is just as vulnerable to mind-numbing routines as a more corporate career.

Despite what we so often assume, committing oneself in service to a corporation requires a great deal of courage. This "in service to" is the key phrase: it means we bring our fullest selves, including our autonomy, to the job. Unlike the grand gesture of quitting to become an entrepreneur, or, on the opposite extreme, giving over one's identity to office politics and corporate dogma, autonomy within a committed corporate job requires quiet strength. Gianpiero Petriglieri, an associate professor of organizational behavior at business school INSEAD and also a trained psychiatrist, believes that this is a skill worth mastering. He points out that it often requires the most courage to maintain independence while also staying the course in business cultures. "It takes tremendous bravery to accept the loss of some control while also maintaining a sense of individuality. You have to work on that every day," he told me.

I worked at Frog Design for seven years, but the first three of those years were spent learning how to navigate the highly complex organization. Only in my last few years did I gain enough political capital to try to have an impact. You might miss out on a tempting new job experience if you commit to an employer for more than a few years, but think of what you gain: institutional knowledge and currency, and most important, a sense of belonging and ownership—a rich, nuanced relationship to an organization that reveals more of its full complexity to you with every passing day.

A conversation with the architect Kay Compton taught me something profound about this kind of commitment. Compton and her husband, who are both avid sailors, made a pact to sail across the Pacific Ocean, the storied "milk run" that hits all the beautiful beaches of the South Pacific. In 2007, they sold their home in Seattle, got rid of all of their possessions, charted a course, and left for a journey that would take them off the grid for the better part of two years. For most of their journey, they encountered calm seas: beautiful sunny days and warm breezy nights.

Sailing alone, Compton and her husband alternated duties: while one served as captain of the "watch," the other stayed below.

"It was a challenge being together like that for twenty-four hours a day. You have to have the ultimate trust in the other individual. Your life is in their hands, and you have to have faith that they will keep you safe," she told me.

Trust and faith were not merely theoretical issues in the middle of the vast, empty ocean. Their boat was headed down the coast of California when they hit their first "big weather," fifty-five-mile-an-hour winds with twenty-five-foot waves.

"I realized that we were at the mercy of our preparation," Compton recalled: "When you're that far out, you can't call anyone. It would have taken the Coast Guard days to find us. You can only rely on yourself and your partner. Once you realize that you can't walk out, you come to terms with it. You make peace with your situation."

That sense of commitment—confronting the given circumstances and finding strategies for working within them—ultimately brought Compton and her husband to safer shores. But she still thinks about those moments of existential fear and reckoning.

"In my daily commute, there is one stoplight where I'm facing the water, and I always think about our journey. I remind myself of what it was like to be out there, to be surrounded by the ocean with just our boat. That reminder keeps me above the fray: it helps me take the long perspective."

When I asked Compton if the experience changed her sense of self at work, she didn't miss a beat before answering. "Absolutely. When my husband and I were on the boat, there could only be one captain. We switched back and forth, but only one of us could lead at a time. This comes back to my work as an architect: you have to work in teams, and you have to understand your role and the roles of others. There is a clarity that needs to be in place with these roles. They can shift, but when someone is the captain, you don't argue. You don't start an argument in the middle of the ocean."

Although not all of us will find ourselves working in such extreme circumstances, her words touched me deeply. How many times have I started arguments in the middle of a proverbial ocean, that is, in the middle of a strategy session or campaign? Of course there should be a place for criticism and discussion in business, but there should also be a place for putting faith and trust in the designated leader. Commitment means accepting the mandate and finding the best way to execute it.

"I wish everyone could feel what I felt on that journey," Compton told me. "If they did, the world would be a very different place; I guarantee it."

Of course, such commitments necessitate vulnerability. Too many of us spend our careers safely hedging our bets, standing on the sidelines, waiting to see who will win the game. If we truly want to commit ourselves, we must put our skin in the game.

Romantics know this is no easy feat. In fact, earlier in my career, I did everything possible to avoid being perceived as vulnerable. Whenever I wanted to win people over for a presentation or initiative, my goal was to deliver a bulletproof case. I went into meetings overprepared, heavily scripted, armed with data and details, ready to address even the remotest concern. I bulldozed any hint of doubt with what I thought was an irresistible combination of self-explanatory benefits, obsessive due diligence, and contagious passion.

I left many of these meetings triumphant, believing I had garnered strong support. It took me a few years to realize that I was wrong. There was an important difference between "thumbs-up" and "we're in." I had gotten a lot of thumbs-up, but I had often fallen short of gaining real, lasting commitment. My colleagues backed my ideas, but they didn't really invest themselves in them. They became followers of my idea instead of collectively leading *ours*.

I finally had an epiphany: the perfect pitch is the imperfect pitch. No one wants to just execute someone else's perfect plan. No one wants to add to something that is already complete. We are all looking for an invitation to co-create, co-opt, and co-own; for an opportunity to make something new, no matter how small or insignificant it is in the grand scheme of things. We don't want to answer rhetorical questions or just fill in some blanks in a flawless design. We want to interpret, tinker with, or even "hack" an idea. We want to be able to deviate from the script. We only truly commit when we are a part of the performance, too.

Today, I try to be more mindful of creating space between my idea and how I present it. When I am delivering a presentation, I rarely reveal how advanced it is in my head. I take a few steps back. I edit; I delete. I purposefully omit things. I design gaps for others to fill. I refrain from mitigating all flaws. The plans I present are incomplete by design and still fragile. They have loopholes, bruises, cracks. They are so vulnerable that others see an opportunity to protect them, to strengthen them, to make them their own. And this is the moment when commitment begins in earnest.

May this be an imperative for the Business Romantic: We are not playing to win a zero-sum game. Our scorecard is always unbalanced. There's always something missing in our equation. "Perfection is characterlessness," the artist and music producer Brian Eno once said.[6] In imperfection we form our character—and find romance.

Just as we Business Romantics push to make ourselves vulnerable, we also look for hints of vulnerability in the people we work with. We are not interested in their "likes," we want to know what they *love*. We are convinced that taste, aesthetics, or shared "cultural capital," in sociologist Pierre Bourdieu's coinage,[7] is an important requirement when hiring people. This is why I recently started asking job applicants to fill out the Proust Questionnaire, just to learn more about who they really are, outside of their streamlined profiles and polished résumés. The Proust Questionnaire was named after French writer Marcel Proust, who, as a young man, filled out a questionnaire about his personality for a "confession album" of a friend. It includes questions about "favorite heroines in literature," "favorite qualities in a man/woman," "favorite virtues," "favorite poets," "your idea of happiness/misery," and other personal matters and aesthetic preferences. In 1993, *Vanity Fair* began featuring the questionnaire on the last page of its magazine.

As I used the Proust Questionnaire, it was amazing for me to see how much I learned about these potential colleagues. Some applicants tried to game the questionnaire and be extra clever and witty, but in most cases you could quickly distinguish a false tone from a true passion. What is your own version of the Proust Questionnaire? What is your bible of good taste? What are your colleagues' passions? Make it an imperative to learn about their loves. Show them yours!

There is no love without vulnerability. In his commencement speech at Kenyon College, the writer Jonathan Franzen pointed out that our modern lives are essentially designed for "likeability," which he views as "the commercial culture's substitute for loving."[8] Liking is far more convenient than loving. All it takes to like is a quick click on Facebook; all we need to do is run out to the mall for the next best thing, and we will have achieved an act of "like." If "like" is all about control, then love is the ultimate loss of control. We put ourselves at risk. We suffer. We open our deepest wounds and dreams to the world, and we attempt to make something beautiful with them.

Love is, perhaps, the biggest word, but any emotion at the office is a tall order. Robert Kirby, a literary agent based in London, told me that although he is comfortable using the word "love" around the office, it still feels taboo to use it in a management meeting. "In order to be taken seriously in business," he said, "I feel I should structure my thoughts in a way that sounds more 'serious.' That has statistics and precision—less inspirational and more 2-D." Indeed, as most of us would attest, show (hurt) feelings in a meeting, and your colleagues will interpret it as a weakness to exploit. Admit ignorance or mistakes, present a foolish idea born out of an impulse, and you may risk being ridiculed for "not thinking with your head." In the corner office, at the water cooler, the board meeting, the employee town hall: embarrassment awaits those who are "romantic" about a business issue. But why are we so terrified of embarrassment? "Embarrassment is sometimes a good sign," singer-songwriter Andrew Bird remarked about writing love songs. "It can mean you are revealing something true."[9] Business Romantics have goose bumps, and, yes, they blush. "We need to move away from the idea of the 'businessman' or the

'CEO,' this Dragon's Den, bullying atmosphere, competitive aspect of *The Apprentice*, the psychological threat rising out of fear," Kirby argued.

Sometimes, in the dullest moments of dull meetings, when some of the participants rattle off their list of issues or launch a stinging ad hominem attack, a million different film clips may start rolling in the romantic's head: "What if I suddenly started to cry? Burst into tears? Told them about my secret loves? Would it destroy me? Disarm them?" In those moments we have two options: we can disengage, or we can commit with our full hearts. When we disengage, we lock away our most private desires; we slowly kill any hope of finding the thrill again. Every time we disengage, a small part of our love dies. When we make ourselves vulnerable, however, we choose to risk it all. We bring the full range of our emotions, and we throw them out right there, right here, right now.

The next time someone makes a disparaging remark in a meeting, express that your feelings are hurt (in fact, use that exact language). When you learn that someone made a negative comment about you behind your back, confront them, but admit your own vulnerability instead of scolding them. Tell them that you are committed to making it across the ocean. Together.

As romantics, we must honor our commitments to work in the middle. We can think of ourselves as sailors crossing an ocean or as teammates with our skin in the game. Romantics know that life's riches are only realized when we share a history with something greater than ourselves. This is easy if our work and careers give us satisfaction. But it can be daunting when the middle begins to feel deadening. The romantic meets these challenges with excess: an excess of attention, an excess of vulnerability, and an excess of commitment. As avant-garde composer John Cage once said, "If something is boring after two minutes, try it for four. If still boring, then eight. Then sixteen. Then thirty-two. Eventually one discovers that it is not boring at all."[10] Commit to boredom; commit to service; commit to inhabiting the space between autonomy and devotion. Above all else, commit to the work. For two minutes, for four minutes, for eight, sixteen, thirty-two. For a lifetime.

CHAPTER 11

Take the Long Way Home

> It's a twinge in your heart far more powerful than memory alone. This device isn't a spaceship, it's a time machine. It goes backwards, and forwards . . . it takes us to a place where we ache to go again. It's not called the "wheel," it's called the "carousel." It lets us travel the way a child travels—around and around, and back home again, to a place where we know we are loved.
>
> —DON DRAPER, in *Mad Men*

With each new season of the award-winning TV series *Mad Men* returning to U.S. households, a sense of nostalgia returns as well. "Oh, the sixties"—that glorious decade, when big ideas still mattered, whether it was the first man on the moon or the creative genius of ad men on Madison Avenue. We take exquisite pleasure in the madness of *Mad Men*. Characters like Don Draper lend the domain of advertising a cultural sanctity. At the same time, the men and women of Madison Avenue never seem sanctimonious or boring because the personal chaos they inhabit mirrors the seismic shifts of the period.

Their tension between morality and immorality, stability and anomie, surface and depth is signature romantic. If romantics are anything, we are conflicted. We long for that amber-hued period in the past when work provided a bulwark against all the erupting madness of the world—yet we simultaneously celebrate the moments when fortification fails, exposing fissures across class, gender, and racial lines, revealing to us "the allure of messy lives."[1] Don Draper himself cites a poem by Frank O'Hara describing the contradictions of this yearning: aren't we all "quietly waiting for / the catastrophe of my personality / to seem beautiful again"?[2]

Nostalgia has been an evergreen in movies, too, perhaps most vividly portrayed in *Casablanca* when Humphrey Bogart and Ingrid Bergman reassure each other of their prized, exclusive possession: "We'll always have Paris." Nostalgia is not only a longing for a time gone by, in which all worldly matters were imbued with meaning; it also refers to a more timeless, existential sentiment. Coined by seventeenth-century Swiss physician Johannes Hoffer, who attributed soldiers' mental and physical maladies to their longing to return home, the term "nostalgia" is a combination of the Greek *nostos* (home) and *pain* (algos). Nostalgia means that we're suffering from an "old wound": cut off from a deep connection to profound truths. As knowledge workers, we exchange information for the sake of instant gratification and incremental gains, using technology to optimize our efficiency and productivity. Yet, as romantics, we are nostalgic for the period when the future seemed less predictable and the world less rushed.

No one captures this sentiment better than the "Amish Futurist," the alter ego of Alexa Clay, whom we met in chapter two. Through social technologies, the Amish Futurist wants to spread the "low-tech prophecy," as she calls it, and celebrate "analog nomadism." Alexa's goal is to probe for the "why" in the development of new forms of technology. Why are we making it? What will it really add to our lives? She describes her agenda as Socratic: by taking a moral stance on the role of technology, she can ask these existential questions of the software community. Yet, because she appears at technology conferences dressed in her Amish clothing and speaking in a soft voice (at a recent gathering in Berlin she baffled an audience of 500 digital industry executives with a meditation on "the power of buttermilk"[3]), her inquisitions are never seen as aggressive or off-putting. "People love it," she told me. "They get excited about the experience of talking to a 'real' Amish person." The Amish Futurist is on Twitter, but reluctantly so: "We telegraph our tweets—they are then transcribed by people in India who upload them online. Social media is an abomination."

Novelist Charles Yu laments that he can no longer fall in love with technology and confesses, "The sexier our high-tech stuff gets, the less I am able to feel anything about it."[4] He remembers the excitement of opening his e-mail in-box: "A private channel had opened up, a vast network of channels, connecting the inside of my head with the insides of other heads. And that network became part of my inner cartography. It changed my map of reality. The physical world gained a new dimension, intangible but no less real." For Yu, the problem is that technology has now become too good at representing our real world. He speaks with regret about a nostalgic "possibility space—not in the mathematical sense, but a place inside that screen where, at least in theory, anything could happen."

The traditional technologist's task is to expand the territory of predictability, to map out the truth, and to widen the comfort zone. In contrast, the romantic-as-technologist expands open-endedness and ambiguity, seeking authenticity by widening the *dis*comfort zone. Instead of seeing technology as serving a utilitarian purpose, the romantic engages with it to capture the strange beauty of the world and create a nostalgic sense of wonder.

This sense of wonder is what is being taught at the School of Poetic Computation in New York, an artist-run organization that uses programming to make works of pointless beauty.[5] The majority of the adult students have experience in programming or in design, but a handful of them have come from far afield: a beat boxer is enrolled, as is a PhD candidate in criminal justice. The founders say their intention is to promote work that is strange, impractical, and magical. Their motto is "more poetry, less demo," because "demonstrations are driven by the end goal," whereas poems are made valuable through "aesthetic and emotional impact." First projects include an Eyewriter that allows graffiti artists to draw with their eyes, and a Sonic Wire Sculptor that uses a 3-D drawing tool to create music.

This kind of training reminds us that certain innovations are powerful simply because they remind us of earlier moments in our lives, moments when we reveled in the simple joy of making and creating. We call ours the age of connectivity, but we might well call it the age of reconnection. We are keen to attach ourselves to something intuitive we once knew but have forgotten over time: the integrity of the person we want to be and the one we actually are, the integrity of left brain and right brain, science and art, reason and heart. We befriend old school friends on Facebook; we revisit places on Google Earth we once visited in the real world; we play songs on Spotify that we liked when we were teenagers; we watch movies on iTunes that meant the world to us when we were growing up. This Proustian "remembrance of things past" is multiplied and amplified in the virtual world. The echoes of our lives finally have a chamber. The first friendships, the first kiss, the first love, the first car—we are trying to bring back the innocence of the "first" when everything tasted, smelled, and felt fresh and promising, and life was one big realm of possibility.

Imagine what different twists and turns our lives might have taken had we followed the stranger whose eyes met ours on the subway. We may not be aware of it when we're young, but as we grow older we realize that our unlived lives weigh as heavily, if not heavier, than the ones we lived. "I bear the wounds of all the battles I avoided," the Portuguese author Fernando Pessoa writes in *The Book of Disquiet*.

This "old wound," this nostalgic longing for connections and reconnections, finds expression in today's market trends. It is at the heart of the rise of local artisans; and it also underpins the Maker Movement, promoting the renaissance of traditional arts and crafts, DIY engineering, and the resurgence of hardware, all of which are driven by the quest for a hands-on experience of work that overcomes the alienation between maker and product. Inventiveness, prototyping, and craftsmanship are being celebrated; Maker Faires are booming; and "hackerspaces" have entered academic and corporate campuses.

TechShop, for instance, a chain of "maker spaces," provides members with an opportunity to "build stuff." For a little more than a hundred dollars per month, members have access to open workshop facilities, industrial equipment, and software for building their own product prototypes. Jim Newton, TechShop's founder and chairman, says he is delighted to see the traditional, "absurdly long" pathway from idea to product shortened drastically.[6] He can point to some notable success stories such as the credit-card-swiping device Square that was developed at a TechShop facility. But the main value of TechShop, in Newton's opinion, is that it enables both consumers and corporate employees to collaborate on personal ideas. General Electric and Ford are among those Fortune 500 corporations partnering with the fast-growing franchise, in an effort to propel reverse innovation and invigorate their own manufacturing cultures with more elements of co-creation and open workshopping. For the Maker Movement, everything old is new again.

Although it is an "old wound" that never heals, research has shown that nostalgia can help with transitions, stimulate our generosity toward strangers, and guide us through periods of boredom, anxiety, and loneliness.[7] Perhaps this is why, consciously or not, it is increasingly reflected in our consumer products and experiences. While they appear to be about novelty and the future, the most meaningful of them remind us of a basic human quest: think of the tablet and our desire to touch; or social sites such as Facebook, Twitter, Instagram, and Pinterest that cater to our innate urge to share (and in the case of Instagram, to instantly "age" a photo through filters). These are new products and services designed to take us into the future, and yet they connect us with a certain idea of the past, resurrecting the romance.

Nothing is more nostalgic than the emergence of handwriting on the web. The web service Bond allows users to send handwritten notes to friends.[8] And the website Think Clearly features a number of handwritten agendas, each one outlining a different "existential" business crisis: "This week I felt as though my work was like walking into a sea of uncertainty." The handwritten to-do lists include "Try to imagine when your work is done. When you are no longer needed," and "Write your resignation letter. You are free to write it in any format that you like."[9]

Similarly, the website The Rumpus started charging five dollars monthly for people who wanted to receive old-fashioned letters in the mail from their favorite authors. The company calls it a "print subscription."[10] A traditional experience is reframed through the means of the connected world.

Jonathan Harris, an interactive storytelling artist, plays with different concepts of nostalgia in his work. His Cowbird website allows people to share their personal stories using photos, videos, sound maps, time lines, and casts of characters.[11] The site aims to create long-form narratives that foster a sense of longevity and continuity amid the fragmented social media noise. As Harris puts it: "The Twitter and Facebook stuff seems to be constantly devoured by its own novelty. Every item is smothered by the one that comes after it, and that happens continuously, 24 hours a day. There's no sense of a collection or of building anything. One thing that I've been trying to get back was that feeling of when I used to keep sketchbooks, knowing I could pass down a record of [my] life. It's that feeling of building something, not just drowning in the moment all the time."[12] Users can communicate with each other with the "gift" of their story. Harris said: "It takes a little bit of the loneliness out of our online existence, which can stem from just shouting into the void."

In a related spirit, the online time-travel site FutureMe.org enables people to send e-mails to their "future selves."[13] The service delivers messages to your e-mail address years or even decades after you wrote them (you can determine the exact date), and the delight lies in the surprise of receiving a letter from the past. A collection of anonymous e-mails was released as the book *Dear Future Me.*

The power of nostalgia is also an increasingly effective ingredient in marketing campaigns. For example, the nonprofit CARE installed large-scale exhibits of "care packages" at high-traffic sites across the United States.[14] These care packages harken back to CARE's original name—Cooperative for American Remittances to Europe—and its role in creating the first-ever care packages sent to European survivors after World War II. Each one of the installation packages features an outsize element such as a thriving crop of corn or a verdant village bursting out of the box, conveying the message that CARE is delivering self-sufficiency, economic opportunity, and justice, not mere aid. The organization is reclaiming its rightful ownership of the concept, restoring the package's aura and expanding its reach far beyond its current associations with sleep-away camp and college dorm rooms. The campaign invites us to reimagine those very first packages arriving on the shores of Le Havre in May of 1946, reconnecting an everyday concept with its rich historical roots.

All these products and experiences reveal insights about what I call "retro-innovation": ideas that mimic an experience of the past in order to transport the user back into a bygone era, or use a new format to meet an "old," sentimental need.

Nostalgia can also help explain the rise of the curatorial. One particularly successful example is Maria Popova's Brain Pickings, a "human-powered discovery engine for interestingness."[15] Popova celebrates subjectivity by sending subscribers an assortment of articles chosen simply because she deemed them worth reading. We crave these curatorial figures: modern-day versions of the old-fashioned concierge who acknowledge us as unique individuals, offering up an escape from algorithmic recommendations and filtered preferences.

This same respect for subjectivity can also be applied to customer research. The field of design research purposely treats its subjects as complex human beings that can only be understood through inquiry and close, if not embedded, observation. Elyssa Dole, a design researcher I spoke with, even created "museums" of the customers she studied. She built a little shrine for each of them, decorated with their portraits and housing all of the memorabilia and personal artifacts that she could possibly collect during her research: diary notes, drawings, photographs, printouts of e-mails and text messages, transcripts of interviews, subway tokens, concert tickets, boarding passes, magazine articles, souvenirs, and other items that were more or less significant for evoking the essence, or at least the image, of a person.

This excessive portrait of the customer as a museum appears to be inspired by Turkish writer Orhan Pamuk's novel *The Museum of Innocence*, in which the male protagonist, after a tragic ending to a nine-year-long romantic relationship, goes back and begins to collect every single object related to their love story as it unfolded from start to finish.[16] He eventually exhibits them in the house of the woman he was in love with, converting it into a "Museum of Innocence" (there is now in fact an actual Museum of Innocence in Istanbul based on the book). As Pamuk writes, museums are places where time is transformed into space. They are containers of nostalgia and protect our greatest dreams, desires, and hopes. They show that which we long for and defend the radical idea that another life is possible.

The museum of museums is NASA's Voyager Golden Record, a collection of human sounds and images on board the Voyager satellite in outer space.[17] Every piece was selected to portray the ingenuity and diversity of life on earth, so that an extraterrestrial life-form might understand what it means to be human. It features works of, among others, Beethoven, Guan Pinghu, Mozart, Stravinsky, Blind Willie Johnson, Chuck Berry, and Kesarbai Kerkar. Even if these artifacts are never heard by aliens, they will forever remind future humans of our most inexplicable longings, tethering us to our terrestrial home.

Business Romantics cherish the "old wound"—our longing for a home—as an important source of romance. We create businesses, products, and services that don't fix the pain but legitimize it as a way to reveal and build character. Sites such as Futureme.org and Think Clearly allow us to find solace for our deepest fears about the future. The Maker Movement and DIY sensibility, including TechShops and "hackerspaces," acknowledge our need for the most grounding of experiences: the tactile, the sensuous, the work of the human hand.

Ask yourself: How can you create businesses that reconnect

us with an essential sentiment unfulfilled in our modern lives? Which traditional human experience can you update and bring to life with technologies of the digital age? What is your museum? And what is your foundational myth, the romance at the heart of your story? What is your "Paris"? Where is your home?

CHAPTER 12

Stand Alone, Stand By, Stand Still

> She told me she had plenty of people to do things with, but nobody to do nothing with.
>
> —PETER B. BACH, *The Day I Started Lying to Ruth*

On September 20, 2013, viewers of Conan O'Brien's late-night television show were privy to an unexpectedly raw moment: the comedian Louis C.K. went off script and wandered into an existential monologue about our alienation in the universe:

> You need to build an ability to just be yourself and not be doing something. That's what the phones are taking away, is the ability to just sit there. That's being a person. Because underneath everything in your life there is that thing, that empty—forever empty. That knowledge that it's all for nothing and that you're alone. It's down there.[1]

Louis C.K.'s monologue was both hilarious and moving. It touched on the ways that constant connectivity through mediated devices and social media can leave us feeling more isolated and even depressed. Recent social genomics studies suggest that this digital overload may even be diminishing our evolutionary capacity to connect with others.[2] Playing on the title of Alvin Toffler's 1970 book *Future Shock*,[3] the cultural critic Douglas Rushkoff calls our current state of mind *Present Shock*, lamenting "a diminishment of everything that isn't happening right now—and the onslaught of everything that supposedly is."[4] Real-time feedback loops make it harder for us to step outside our echo chambers of status updates and tweets, and be present to our own existential sorrows and joys.

Writer, researcher, and consultant Linda Stone uses the term "continuous partial attention" to describe our "always-on, anywhere, anytime, anyplace behavior that involves an artificial sense of constant crisis."[5] But she proposes that rather than bemoaning our symbiosis with digital devices or making a heroic effort to disconnect entirely from technology, we ought to simply shift the conversation: "We stress about being distracted, needing to focus, and needing to disconnect. What if, instead, we cultivated our capacity for relaxed presence and actually really connected, to each moment and to each other?"

Her vision of connection is medium-agnostic. It doesn't really matter if we are communicating via Skype, texting on our cell phones, or chatting over coffee; her point is that our connections should be based on a genuine investment in the other.

This calls to mind an experiment Priya Parker and her husband, Anand, call "I Am Here" days. One Sunday each month in New York City, a group of eight friends gets together for an outing. Each time, a different member of the group serves as the "curator" and designs an urban exploration. The goal is to visit different parts of New York City while experiencing time and companionship in a more focused and deliberate way. On one occasion, the group headed up to Harlem for a day of discovering historic boulevards and unexpected alleyways. Another outing was designated Red Hook Day, as the participants set out to explore the industrial and rapidly gentrifying neighborhoods along Brooklyn's shores.

The underlying principle for "I Am Here" is the idea that spending eight hours with eight people every month creates deeper connections than spending one hour with each of them eight times a month. The participants are obliged to "show up" and bring their entire self: body and mind. What they don't bring, however, are mobile phones, laptops, and tablets. Referencing anthropologist Clifford Geertz's term for the thickness of contextual data—"thick description"—everybody makes a pledge to being "thickly in one place, not thinly everywhere."[6] As Anand described it, "All day long we talk. We walk, we talk; we eat, we talk; we browse, we talk. It can feel, in the days we live in, like a subversive act."[7] When the conversation organically dwindles, they simply space out or gaze up at the sky.

This idea of thick presence is also the principle behind "cloud spotting." Gavin Pretor-Pinney, founder of the Cloud Appreciation Society, which has more than thirty thousand members, observes the movement of clouds and derives great joy from interpreting cloud formations, "the patron goddesses of idle fellows," as Greek playwright Aristophanes called them. The cloud spotter's endeavor is the perfect case study for how seemingly mindless pursuits can provide opportunities for mindfulness. Pretor-Pinney heralds the immense value of "aimless activity" that leads to absolutely nothing.[8] Cloud spotting doesn't have a purpose and won't help solve any of the world's problems—but that is precisely the point. Clouds epitomize the wonder in the mundane. "You don't need to rush off from the familiar, across the world, to be surprised," says Pretor-Pinney. "You just need to step outside."

Thick presence is slowly becoming a new hallmark of luxury, and even a new social etiquette. In Sydney, some retailers have begun refusing to serve customers who are talking on their mobile phones while they're in the shop. They find it rude and ask for more respect. In many of the city's stores, signs have been placed to remind customers to leave their phones in their pockets.[9]

On the less punitive side, hotels and resorts are experimenting with giving their guests the experience of thick presence by taking away their devices. Hotel guests pay for the pleasure of relinquishing control. At the Four Seasons in Costa Rica, the "Disconnect to Reconnect" program offers to lock up guests' iPhones for a minimum of twenty-four hours in a safe deposit box.[10] It seems the most connected of us are so incapable of unplugging on our own that we need to pay an outside arbiter to take charge.

Certain businesses have even managed to carve out a customer service strategy based on thick presence. Online retailer Zappos, for example, does not reward the number of customer service calls its representatives complete every hour. Instead, the company rewards representatives for staying on the phone *longer*, coupled with a "happiness rating."[11] Zappos communicates to its employees, and to its customers, that the quality of the conversation is what matters, not how quickly and efficiently it is completed. It places value on being thickly present with each individual and not thinly present across an abstract data set.

All these strategies stand in stark contrast to the expectations of today's leadership, especially in the face of advice that tells us "Inaction is no longer acceptable"[12] and encourages us to conduct "Business at the Speed of Now"[13] and to "Never Eat Alone."[14] If there is one thing leaders across sectors and industries have in common, it is their bias toward action. "What's the outcome?" the invisible hand of the marketplace seems to whisper into our ears, constantly. We connect to lead, and we lead to act.

However, many executives I know secretly admit that their

action bias, their tendency toward swift decisions and driving change, is merely an inherent aspect of their role and the product of a deep-seated insecurity (they must be expected to do *something* and make changes—why else would they have been hired?). Often, it seems, this leads them to take Joseph Schumpeter's macroeconomic concept of "creative destruction"—the idea that innovation emerges from the incessant mechanism of destroying the previous economic order—and apply it to their own teams and organizations.

In our Western business culture, we generally favor leaning in, decisiveness, and speed over contemplation and conversation. Value is placed on shorter "time-to-market," and "lean" is better than "thick." Ideas are deemed irrelevant if they cannot be implemented quickly; insights that are not "actionable" are considered inconsequential. Business lingo values "getting things done," material output, and busyness over the apparent inefficiency of discourse and discussion.

It is not surprising that in this intellectual climate, business conferences are often dismissed as mere "talk shops." The idea that talking and exchanging ideas alone may have inherent value seems to be losing currency. Conferences such as the World Economic Forum, TED, PopTech, and others have therefore become more focused on producing tangible outcomes. "Impact" is the magic word. They have launched incubators, "accelerators," and award programs to reward the doers among the thinkers; and they have slowly morphed into think-*and-do*-tanks. Only the cultural sector continues to appreciate unconditionally the merits of spirited exchange. It needn't be as excessive as the twenty-four-hour interview marathons of the Serpentine Gallery in London[15]—prime examples of thick presence—but the idea of creating generous spaces for indecisiveness and reflection, of halting time for countless hours of just listening, can be invaluable to business leaders, too. Talk must not be cheap.

J. Keith Murnighan, a professor at Kellogg business school and author of the book *Do Nothing!*, was one of the first management thinkers flagging the risks of "overmanaging" and preaching the gospel of management minimalism.[16] Lately, he has received more support from mainstream voices. Even the *Economist*, an unlikely practitioner of "slackerdom," is now praising the virtues of "laziness" and "wait and see,"[17] pointing to research showing that soccer goalkeepers who remain right in the middle of the goal—avoiding extreme choices to the far left or the far right—stand the best chance at catching penalty kicks.[18]

This kind of passivity demands that leaders cultivate the ability to actively let go. Inaction, leaning back, and expanding our time frame beyond quarterly reports and annual earnings requires courage and discipline. Thinking is a solitary act, and it cannot be expedited by browsing the web for the next big thing, studying the competition, telling others what to do and then monitoring them, or crowding calendars with back-to-back meetings. It necessitates an independent, a lonely mind.

In his commencement speech on "Solitude and Leadership" at the U.S. Military Academy at West Point, the writer William Deresiewicz remarked that "If you want others to follow, learn to be alone with your thoughts":

> It seems to me that solitude is the very essence of leadership. The position of the leader is ultimately an intensely solitary, even intensely lonely one. However many people you may consult, you are the one who has to make the hard decisions. And at such moments, all you really have is yourself.[19]

In the business world, we have fallen prey to our own special version of a new atheism, touting aphorisms such as "hope is not a strategy" or "belief alone is not enough." In business, we do things for a reason, and we spend the majority of our time trying to act rationally, teasing out contingencies, assessing risks, analyzing and cross-checking every possible scenario before we act. We mistrust people who simply follow their intuition; we cannot fathom engaging in a joyful but aimless activity such as cloud-spotting; and we consider any pause a setback. We prefer to make decisions. We are always moving forward.

Movement is what spins it all together: cultures, organizations, experiences, and relationships. But it is the counter-movement—doing things for no reason, more slowly than usual, against the grain, or not at all—that grants us romance. And that, in the end, is the difference between a life that is merely productive and one that is truly meaningful. Business Romantics need time and take it. Time for pause. Time to be quiet. Time to find the signal amid all the noise. Time to *be* the signal. We under-manage. We conduct business at our own speed. We enjoy eating alone. We do nothing.

Business Romantics, stand still. Others, come closer.

PART THREE

Into the Fire

CHAPTER 13

Measures of Success

> You can't put being in love on a scale. Either you are or you aren't.
>
> —JENNY HAN, *We'll Always Have Summer*

Now that we have come this far, I hope that you are inspired by the stories and possibilities I have shared. Throughout the Rules of Enchantment I offered you a set of tactics and tools to help you navigate your journey as a Business Romantic. It is time to return to your life's work and your work in life, and fall in love with it all (again).

Inevitably, however, you will face naysayers, skeptics, even mockery and scorn. Anyone who generally finds the notion of marrying romance with business ridiculous will raise a myriad of objections. And, conversely, some of the toughest questions will come from the very people and communities that are most passionate about romance and intrigued by the idea of romance in and through business.

In this final chapter, I play devil's advocate, addressing possible reservations and outright rejections to the Business Romantic concept. I look at the measures of success and failure through the eyes of the romantic, arguing that we should bring Business Romance to scale. This will require some of us to become corporate change makers, others to become public advocates for intimacy, and all of us to become champions and protagonists of a new romantic age. I leave you, at the end, with an Appendix that helps you embark on your own Business Romantic mission: the Business Romantic Starter Kit.

But let's begin with the most immediate questions: How does a Business Romantic approach account for traditional notions of performance? How does it all add up? What does "success look like"? And what do we do when we fail?

Typically, we define success as a progression of achievements, a straight line to the highest plateau of lifelong accomplishment recognized and rewarded at work, at home, and by society at large. In other words: a career. In the conventional definition of a career, success breeds more success, or at least we hope! It's like the monster that cannot be satiated: the minute we succeed, we feel compelled to garner more success, the goal of which is to become even more successful. We know it is misguided, and yet we cannot help ourselves; we are trapped on the proverbial "hamster wheel," constantly trying to spin farther and faster. Writer George Saunders cautions us against the distractions caused by this narrow definition of success: "There's the very real danger that 'succeeding' will take up your whole life, while the big questions go untended."[1]

As I argued in the Rules of Enchantment, Business Romantics design their own endnotes. We create rituals to celebrate and sanctify some of those rounds on the hamster wheel. At times, we even step off, if only for a moment of reflection or a period of rejuvenation. When we leave, we do so with a full heart.

Maybe, like Andrew Mason of Groupon or Frank Rijkaard of Barça, we leave generously and with humility. Or maybe, like the employees at Naked Wines who pressed Send together on their final e-mail, we leave as a collective in search of a more fulfilling life.

As Business Romantics, we recognize that success is a necessary part of our career equation, but we view it with a healthy dose of irony and playfulness. Instead of holding ourselves up to quantitative metrics of success, we have other, more provocative forms of assessment. We wear a mask! We delight in trying out the persona of the high achiever. But we never forget that there are countless other aspects to our souls: vulnerability, melancholy, suffering, and a deep appreciation for strangers and strangeness, just to name a few.

Although cynics may decry the romantic disposition as a comfortable retreat, a "safe haven for slackers," as one of my more skeptical colleagues called it, Business Romantics have no tolerance for self-indulgence. We don't tune out; we zoom in. Romance in business requires conviction, rigor, and a zealous attention to detail. Remember Scott Friesen, the former Best Buy manager, from chapter two? He was driven primarily by the desire for excellence; he told me that his greatest moments in business occur when his entire team is performing at the top of its game, "like a great band that has been playing for years together." This is excellence not only for the sake of the bottom line or a glowing performance review but also for the sake of a transcendent experience, a heartfelt obligation to quality.

You will certainly have to answer to numerous measures of financial and management success throughout your career—and yet, never let go of your own vision for romantic success! What brings you to life? What are your romantic moments? How do you evaluate them? This is no easy task: business leaders need to stay mindful of prioritizing values that don't directly touch the revenue stream. It's a constant battle to fight for time and resources for initiatives that cannot be measured with a set of metrics. Matthew Stinchcomb, of Etsy, spoke to this conflict in his own career:

"I try to find a middle way between the heart and the head, and I hope I always do that. But I'll never be able to quantify it. How many hearts and minds have I won this month? I'll never know." The consequence of following a more romantic vision of business is that the measures of success will be more "esoteric," as Stinchcomb called it: "In certain kinds of business, you can say, 'I built this big thing, and it has resulted in X.' In my role, I say, 'I spent my time giving employees the means and desire to follow the Etsy values.' Am I successful? It's a little bit difficult for me to say. I measure my own success by self-assessment: 'Did I say I was going to do it? And then did I do it?' "

What Stinchcomb describes here is his own mission, the moral purpose for his being in business. In our performance reviews, we are often evaluated against our company's mission—did we live up to it?—but we rarely ask how the company lived up to ours. Organizations go to great lengths to craft a compelling mission statement (for example, "To organize the world's information and make it universally accessible and useful") while we as individuals never articulate our own, maybe because we fear that making it explicit might compromise its truth.

I would encourage you to write yours. Maybe it's just a fortune-cookie statement, a verse from a poem, or something someone said to you one day, or maybe it is more elaborate. In any case, it is a helpful device for affirming your place in the world and describing a unique talent or perspective only you can give. Lofty enough not to be mistaken for a career plan, this mission statement describes your cause worth fighting for. Here is mine:

> With great vulnerability, contagious excitement, and a religious belief in details, I create and defend spaces for possibility and action to demonstrate that, perhaps, we are connected and that there is more than we can see.

This statement serves as a lens through which I look at all my professional (and personal) decisions. It is a compass, not a dashboard. It captures who I want to be, why I'm here, what I would like to do, and how I would like to operate. Every word is the result of careful consideration, and the most important one, perhaps, is "perhaps." As romantics, we can be full of conviction, but never fully certain.

Matthew Stinchcomb laughed with me when we discussed this very issue during our conversation. "The biggest consequence of using my own assessments as a measure of success is my own insecurity and fear," he said. As romantics, we must not try to contain these moments of fear. Our "wins" might not look like other wins, and our "successes" may be invisible to our colleagues and teammates. And, inevitably, there will be moments when we lose. We will lose pitches, clients, market share, and we will even lose face. No one can stay at the top forever. That is why accomplishments, accolades, or the mere perception of success fill us romantics with nervous anxiety. The moment we receive the greatest applause is the moment of our greatest sadness. We are predisposed to such extreme sentiments; we see the dark when there is light and see the light when it is pitch dark.

In these moments, we turn to our imagination. We recast loss as an artful stab at meaning. We don't measure ourselves through moments of defeat. Rather, through reflective questions such as: Did we perform our jobs with honor? Did we strike a chord that reverberates every day? Were we able to find a sympathetic community in the marketplace? Did we get in touch not only with our needs but our wants? Were we true to our full selves from beginning to end? If the answer for us is "yes," then we've lost nothing at all.

Considering all this, is there any real way to measure romance in explicit terms?

No: Defining romance through metrics would require a Borgesian brand book for life, a tome of thousands and thousands of pages outlining every expression, regulating and assessing each individual experience. And yet, with the approach of the ultimate bottom line and with all resources spent, we would still be unsure of the return. Was it worth it? Measuring romance is akin to measuring the value of our own lives. Our final hours before we die belong to our unquantified self. We don't seek an accounting equation or scorecard to evaluate what our life was worth; we can only hope it was worth living. When we reflect on our best days, all of us end up romanticizing, all of us become romantics.

So you still want to see an actual Business Romantic Performance Review? How about this?

> *Dear Joe,*
>
> *Ever since you walked through the door to my office for the first time, I've been a fan of yours. I instantly saw how passionate you are about what you do. You didn't come to play it safe. You put everything on the line, and your heart is always in it. You really care.*
>
> *When you call a 1:1, I know I have to raise my game. Anything can happen. But I always feel safe. I trust you.*
>
> *Everyone who walks over to your desk always gets a quick chat out of it, or at least a smile. I know this sounds cheesy, but it makes this a better place.*
>
> *Your plans are thorough but never perfect, so others can shape them.*

Your presentations are simple yet elegant; they always lead us to new places we didn't know about.

Your memos to the team are a pleasure to read, even if they convey difficult news. I printed out the last one you sent, about our revised strategy framework, and read it three times in a row on a Sunday morning over coffee. I appreciate that you still take the time to sign e-mails with your full name and that you obviously pay attention to punctuation.

I love how you fall in love with ideas, fight for them, and protect them like a shepherd tends to his sheep or a gardener tends to his rosebushes.

You are not afraid to be an amateur, and I still recall the many moments when you said, "Let's do it." And you did, although you had no idea how.

We suffered through so many meetings together like one suffers through French art-house movies. You always stay until the end. Because you know it's worth it.

Remember when we had a postmortem after the launch of our China office? We just sat there for ten minutes, exhausted, quiet, but with our eyes glowing.

Areas of improvement? We want more of you, not less. Don't streamline yourself or spread yourself too thin. Don't hold back. Don't pretend you have a strategy. Don't spend too much time on the marketing plan. It will be obsolete tomorrow anyway. Don't waste your precious hours on dashboards or reports that are supposed to prove your value.

No dashboard in the world can capture the impact you've had.

On behalf of our colleagues and customers: it is a delight to be in business with you.

In true romantic tradition, up to this point, the principles and tools I have offered for more romance in business have been tailored to individuals. This makes sense because when we try to take romance beyond the personal realm, we jeopardize it. Like happiness or any other elusive concept, romance dissipates the very moment we attempt to catch it. It is the proverbial butterfly that alights on our shoulder for a fleeting moment, unnoticed. Try to scale romance and you risk losing it. If everything is romantic, nothing is romantic.

And yet philosophers, poets, and artists did "scale" romance in the nineteenth century. They transformed their individual desires and expressions into a collective movement that spanned the globe. Business Romantics can do the same. Like the League of Intrapreneurs, artists-in-residence, Rebels at Work, or the U.S. Army's Red Team University, we can cultivate constructive forms of opposition. Like Anthon Berg, Reddit, and REI, we can initiate with the design principle of generosity and work backward toward profit. Like McKinsey, we can leave everyone guessing. Like NBBJ, we can create a secret society as a catalyst for organizational transformation. And like MMM, Muji, or Moleskine, we can bring "negative space to life." All of these examples, as well as so many of the others outlined in the Rules of Enchantment, require leadership. Of course, given how much we already know about romance, it should come as no surprise that the Business Romantic leader is decidedly idiosyncratic.

Whether as executives, intrapreneurs, or corporate rebels, Business Romantics are uniquely positioned to lead and facilitate change within their organizations. By incorporating ideas outlined in this book, we can engender an "institutionalized" romance. When we take all those meetings, dinners, walks, projects, and customer experiences—all those best practices—to scale, we begin to build the foundation of the Business Romantic Company.

No, I don't think you need to hire a chief Business Romantic officer (or give yourself that title). However, if you are in the position to hire and manage, this responsibility is obviously a powerful means for instilling and nurturing a romantic spirit in your organization. It doesn't imply that you need to form a Department of Business Romance. And there is no need to create the position of, say, a senior manager of mystique, a director of nostalgia, or a VP of suffering. Nor do you need to hire a hermit-in-residence or a personal clown—as the CEO of Cirque du Soleil did—to create small moments of delightful strangeness. But you do need a whole new kind of "T-Shaped People." Let me explain.

Design thinkers often use the analogy of the "T-Shaped Person" to describe the type of multidisciplinary thinker and doer who thrives in an innovative environment: rooted in one core expertise or discipline, with each of his arms stretching into adjacent disciplines.[2] In contrast, picture the typical Business Romantic more like an expressive pose: one foot on the ground, one arm stretching into unknown territory, the other reaching up into the sky to connect with something transcendent, and one foot off the ground in a celebration of instability, ready to lose control at any time. Try to keep balance! The Business Romantic is uneven, even shaky, volatile, never fully aligned, but always ready to jump.

At the same time, you don't want a team that is composed only of this one type of person. Uniformity is poison to the romantic spirit. This is what I recommend for your Business Romantic Dream Team:

- an additional committed Business Romantic, a trusted confidant, the second-romantic-in-command, so to speak, who is on exactly the same wavelength as you;
- an "outsider on the inside," the rebel, the renegade, the counterromantic (in fact, this may be one of the few meaningful roles for the cynic: the cynic-in-residence on a romantic team) who can challenge you and create the opposition crucial for any true romance;

- a Business Romantic project manager who understands projects as romantic enterprises and can stage them as a series of dramatic events with interchanging phases of hope, euphoria, thick presence, and grief;
- a Business Romantic executive assistant or administrator who views the team's calendar as a narrative thread as well as a daily and hourly reminder of the poignancy of passing time. Supporting the Business Romantic executives, the Business Romantic assistant knows that his or her main responsibility is to make them feel good, rather than making them merely more productive. Keenly aware of all the activities and meetings they could have accepted but ultimately declined, he or she is the keeper of their other life—the life that exists only in possibility. The assistant and the executives he or she supports are intimate strangers, and the calendar is an organizer and discipliner, but also a record of broken promises. It is the modern work life's container of "lost time," which, in today's world, must be retrieved by digging deep below the veneer of iCal—more easily found, but also more easily lost than ever before.

Naturally, you will be closer to some of your colleagues than to others; some of them will meet your spouse and children; some might see you cry or lose your temper in a meeting; some of them will admire you for your leadership wisdom and want to spend more time with you; others will keep a professional distance. This is all as it should be. Most important is that each member of your team provokes a strong reaction from you. You should view each and every one of them as a potential friend or foe. Life is too short to be wasted on milquetoast hires. Neutrality and impartiality are the death of romance, so surround yourself with a team of people who, by turns, unsettle you, inspire you, and challenge you. The last thing you want is impassivity at the table, a group defined by mere "professionalism."

To make the right selections, I recommend using the Proust Questionnaire for interviewing new hires. And then, once you have your Business Romantic Dream Team in place, employ the art of imperfection in all of your interactions! Leave some space in your initiatives and projects so your team can fill in the cracks with its own ideas. And encourage each colleague to do a role swap one day out of every month.

A more conventional business approach would speak of a mix of Why, What, and How people. The Why person is the one who articulates and embodies the purpose, the raison d'être of the team: Why are we all here? Why do we all get up in the morning? Why do we strive together? He or she is the inspirational voice, the moral authority that can craft a vision, establish and hold up the standards of integrity, and remind everyone of the values they share and the mission that unites them. In contrast, the What person is the creator, a strategist who can translate the mission into reality. The How person, finally, is the implementer, obsessing over details and insisting on operational rigor.

A romantic leader is not content with embodying just one of these three types. The Business Romantic shifts between the personas of Why, What, and How from day to day, blurring the lines between strategy and tactics. He tries on multiple identities, wears multiple "hats" (or masks), and performs as "always a different person"—a stranger in a strange land. For the organization as a whole, this means a more unorthodox hierarchy, with a different understanding of roles and responsibilities. Traditionally, organizations have split the Why and What and How among different tiers and team members. Usually, it's a top-down vector—the more tactical you are (the more you are concerned with the How), the lower you rank; whereas the strategists, the Why and What people, rank higher up and hold the more prestigious titles.

Not in romantic organizations: unlike a more conventional business, the Business Romantic Company is made up of a collective of individuals who are Why, What, and How people at the same time, moving fluidly between the worlds of ideas and implementation.

So what does the Business Romantic C-suite look like? The Business Romantic CEO is a searcher, envisioner, and connector: the person at the table with the biggest heart; the Business Romantic CMO is an artist who can win hearts (and, if needed, also minds), using seductive techniques to do so; the Business Romantic CFO cherishes the inherent beauty of spreadsheets and regularly breaks the heart of the CMO; the Business Romantic COO tries to minimize unpredictability (but secretly loves it); the head of sales is driven by permanent unfulfillment; and the head of human resources is the master of ceremonies overseeing how all romantics fall in love and break up. Some of these executives are in the business of mitigating risk; others are seeking it. But all of them are united in their love for business as one of humanity's greatest adventures. They try hard to apply reason to the fruitful chaos of their organization and markets, but they are smart enough to know that at the end of the day, they can't. Instead, they are sailing the ocean together, cultivating trust on a tiny schooner while navigating a vast body of unknowns.

With these teams and leadership in place, the Business Romantic Company creates mysterious customer and employee experiences; honors ambiguity; designs for friction; and provides "critical events." It makes us wonder, and it makes us wait and suffer. It looks for revelations, not just results; it renews the beginnings and celebrates the endings while committing to romance in "the middle."

Throughout the book we have seen examples of companies that excel at particular romantic practices: start-ups and entrepreneurs such as HICKIES that have a romantic foundational story embedded in their DNA; Eileen Fisher, McKinsey, MMM, Muji, Moleskine, Secret, House of Genius, and Secret Cinema, whose business models or organizational designs leverage the power of secrecy, obscurity, and absence; companies such as Somewhere or the Barbarian Group that celebrate the office as a space for social exchange and storytelling. And then, inevitably, there is Apple, with its appreciation for beauty and the spirit of things (famously, Steve Jobs cared a great deal about the interior design of Apple devices). All of these companies share a contrarian, rebellious spirit, as well as unreasonable behavior, a tribal culture, and the ability to be foolish without ever compromising their obsessively high standards of excellence. Because of their exclusivity, pride, and ambition, these romantic organizations are rich with conflict and opposition, whether in the form of heated debate and politics or in the design of opacity and ambiguity. They are high-risk, high-reward places, full of danger and delight. They are hotbeds of human drama, simply because the dreams they dream are so big, so much is at stake, and so many have skin in the game, far beyond just financial success.

Whether you work for a for-profit, nonprofit, start-up, small or medium-size enterprise, or Fortune 500 corporation; for a "conscious capitalism" company or a more traditional member of the business mainstream; in human resources, marketing, R&D, sales, or finance; as CEO, administrative assistant, project manager, or "director of first impressions"—you can do your part to make your company a romantic one. Your success depends on how actively you can apply the Rules of Enchantment. It depends on how freely you can create your own romantic sub- and countercultures at work. Tell secrets; close doors; be excessive in your pursuit of excellence; start off meetings by saying something ambiguous; end on a percussive note; identify the "misfits" and honor them; wear a mask; call attention to your own vulnerability; fake it! It all depends on your ability to be unreasonable, to appreciate the many returns that are "beside the point." And more than anything, it is predicated on your willingness to give everything, to be in it and of it. Lose yourself in your work. Come back to yourself as a different person.

CHAPTER 14

The New Romantic Age

To begin, begin.

—WILLIAM WORDSWORTH

At this point, even those of you who identify as romantics might still balk at my attempt to portray business as the great re-enchanter, the great signifier of our lives. It is too much, you say, for business to captivate the marketplace with romance. It gives business even more power to permeate our societies. It will further erode the privacy and integrity of our lives, you insist. A marriage with the market system is the ultimate betrayal of romanticism!

These are legitimate concerns, and the risks are real. Market values have invaded all aspects of our lives. On the one hand, this has provided us with an efficient mechanism for making decisions and solving problems, and for determining the value of ownership and access. On the other hand, however, it has also commodified our behaviors, our relationships, and even our moral judgments. In his book *What Money Can't Buy: The Moral Limits of Markets*, the moral philosopher Michael J. Sandel observes that "putting a price on the good things in life can corrupt them." Sandel argues that markets change the character of the goods they trade, a phenomenon he calls the "expressive effect" of markets. Think of the Olympic flame. If I had taken it home as an exhibit while I worked for the relay, charging fees for access to it, what would that have done to its meaning? Or think of relationships. Friendships may ultimately follow a tit-for-tat logic, with reciprocal benefits as underlying catalyst, but we chafe when explicit market mechanisms are used to negotiate our most intimate interactions. We would never pay our friends to listen to our troubles; we don't submit a performance review to our parents after they watch our children.

Arlie Russell Hochschild, author of *The Outsourced Self*, contends that there is danger in using the market to delegate these fundamental relationships to an external party.[1] Instead of having friends, we now pay a therapist; instead of relying on our relatives to raise our children—"It takes a village"—we now employ nannies, sleep consultants, and baby concierge services; and instead of asking our neighbors for a favor, we now sign onto TaskRabbit for our weekly errands. The efficiencies of market mechanisms, Hochschild argues, are not an appropriate substitute for the muddier and less baldly transactional aspects of our personal selves.

As a Business Romantic, I couldn't agree more. I would never suggest that we attempt to outsource empathy, passion, commitment, and dedication to a third party. What I do argue, however, is that market mechanisms can offer up surprising moments of kindness, delight, and intimacy. Think of Suspended Coffee and the Generous Store; or the Harteau family connecting with strangers through its 24-Hour Bazaar; remember my own surprise and gratitude when my wallet was returned, or the viral success of the First Kiss video ad. Just as personal interactions often have hidden transactions embedded in them, business transactions can be vehicles for meaningful connection.

Still, critics wonder if business has the nuance to embrace all of the contradictions inherent in a romantic perspective. After all, a romantic point of view is not always a happy one, as any romantic novel from the nineteenth century will clearly illustrate. Like any humanist movement, romanticism demands that we acknowledge the full complexity of our character. Gianpiero Petriglieri, the INSEAD professor, is skeptical that the business domain—usually associated with positivist buzzwords—can handle romantic modes of ambiguity, conflict, and drama. To him, the idea of Business Romance is problematic. "If romanticism is an experience that values the primacy of impulse over rational consideration, our wish to give sovereignty to our own impulses and experiences might better remain in the private sphere, away from business and commerce," he told me.

Other critics may argue that the idea of Business Romance is just a luxury of the developed world, a sign of decadence in which we can indulge once our more basic needs are met. They may mock romance as a kind of yoga for the workplace—a new lifestyle drug that covers up the tough economic challenges of income inequality, the yawning digital divide, and structural unemployment that now loom over our global societies. Romance, they argue, is just an aspirational privilege for the lucky one percent. It is not something truly fundamental and obtainable for everybody.

I take these and other critiques seriously. Romance may describe some basic human sentiments, but even the idea of romantic love as the basis for marriage—rather than economic motives—is a fairly recent phenomenon. It wasn't until 1943, when Abraham Maslow developed his famous hierarchy of needs, that such a form of expression was prioritized in society-at-large.[2] If you're working in a service job for minimum wage, material gains will make a much bigger—if not the *only*—difference to your well-being. In such a case, cash flow might matter more than adrenaline rush. For an increasingly large group of people living in the United States, there has never been much romance in work, and growing inequality may dash hopes for any such romance in the future. In light of this, romance is indeed a privilege, and one that is more easily obtainable for knowledge workers in stable economic conditions (certainly, the majority of individuals featured in this book qualify as such) or professionals in other sectors who do not have to fight for survival every day. It is a luxury to be able to ask "why?"

Business Romance doesn't necessarily mean that you "do what you love." In fact, this Steve Jobs–inspired mantra so common today among today's micropreneurs, local artisans, and free agents is starting to receive some pushback. Writer Miya Tokumitsu argues that such an adage trades in elitist assumptions which degrade the inherent value of work and duty.[3] Business Romance also doesn't mean that you always "love what you do." What it does mean is simply the ability to create and find *moments* of love in what you do and how you do it, even in what Tokumitsu refers to as "un-lovable" labor: conditions for genuine human connections, a yearning for the faintest sensation of greatness transcending the mundane. You may call this an escapist manual for self-deceit: I call it the courage to fight "death by realism."

My hope is that a taste of romance may serve as a point of inspiration for everyone: for those living in advanced consumerist societies in which the market economy has commoditized our relationships, compartmentalized our selves, and secularized our lives; and those in which the daily business of life is survival. Romance can offer us a beacon of hope and a transcendent glimmer in times of ruthless logic, resignation, and depression. The quest for romance may not be a salve to all of our wounds, but it can be a compelling coping mechanism. This is not a bad thing. As romantics, we are escape artists; we dream up, create, and defend the conditions for the possibility of another, a better, life.

Yes: we could easily untangle all of the ways our free market system corrupts the meaning of goods. And yes: the language of management might sanitize our darker notions of romanticism. But it is too late to put the genie back in the bottle. We cannot achieve a pre-market-society innocence. What we can do is use markets to tell new stories, ones that better reflect our inclinations. Our romantic "infinite longing," as the writer E.T.A. Hoffmann called it, will never be fully satisfied by the finite resources traded in markets. Yet the marketplace can give us opportunities to connect, to alleviate our loneliness and isolation, to affirm that our private longings are, in fact, a collective desire.

As Business Romantics, we can choose to remain private, protect our flames, and keep them hidden from scrutiny, criticism, or scorn. Romance is impossible without such privacy. As my former Frog Design colleague Jan Chipchase, an ethnographer, once wrote, "Only people with dull lives can afford to forego privacy."[4] And yet, the pursuit of Business Romance is impossible without the messy vitality of public life, without full engagement with the world. Publicity means somebody else is always there with us. For romantics, this is reassuring. Like artists, we need an audience, real or imagined, worldly or otherworldly, or our enterprise deflates. If nobody is watching us, why bother? This is why we must ultimately expand the scope of our romantic perspective. We must start with ourselves, bring romance to our companies, and then, finally, romanticize the world.

Romance has always been both a deeply private and deeply political position. Think back to the original Romantic Movement described earlier in the book. Its conflicted relationship to a public and private life serves as a lesson to us today, revealing that something vital is at stake, not just for business, but for society. In the U.K. of the 1790s, for example, the writing of romantics was considered an affront to the supremacy of reason and usefulness. The romantic poets William Wordsworth and John Keats were even put under state censure by prime minister William Pitt.[5]

This oppressive side of the Industrial Revolution was made perfectly manifest by philosopher and legal reformer Jeremy Bentham's 1791 concept of the "Panopticon": a cutting-edge innovation in the design of institutional buildings that allowed one single inspector to observe large numbers of the building population (for example, inmates, workers, patients). Bentham—considered the founding father of utilitarianism, which attempted to maximize the greatest amount of happiness for the greatest number of people—envisioned a utopian universe of transparency. In his fantasy, the entire world would exist as a series of spaces and places in which "each gesture, every turn of limb or feature, in those whose motions have a visible impact on the general happiness, will be noticed and marked down."[6]

Bentham's drawings were never utilized in an actual construction, so he would surely be delighted to see how fully realized his "utopia" of complete and total surveillance has become today. More than two hundred years later, we find ourselves voluntarily carrying tiny versions of Bentham's Panopticon in our pockets. And smart phones are only the beginning. The recent revelations about government surveillance on a massive scale seem like an eerie déjà vu: "A system of spies is unraveling society's beautiful fabric of love," the romantic poet Coleridge lamented in 1795.[7]

Today, we are no longer under surveillance by someone else; we are under surveillance by ourselves. We are constantly watching and recording our lives, offering up massive amounts of personal data—our Quantified Selves—to an omniscient technocratic deity, pursuing total visibility in the name of utilitarianism's greatest happiness principle. While the Panopticon allowed the few to see the many, hyperconnectivity now also enables many to see many and many to see few. We are becoming prisoners of the technologies that are tracking us, filtering information for us, and, ultimately, deciding for us. Our new Panopticon might be Google Glass, and we might all become consumers and citizens of "glass," fully transparent to everyone at any time.

The new currency of our market society is personal data. Today businesses and governments collect, analyze, and exploit ever greater quantities of it, monitoring individuals through wireless social sensing technologies and satellite imagery.[8] Technology critic Evgeny Morozov argues that datafication enables democracies to amass so much knowledge about their citizens that they can engineer "a perfect, highly personalized, irresistible nudge to their self-interests."[9] Perhaps, worse, as our personal data turns into an asset chamber to exploit, what was formerly public domain is now being governed by a powerful group of technology companies solving problems: "algorithmic regulation."[10] This trend is most prevalent in the high-tech innovation circles of Silicon Valley where "alpha-geeks," the glorified start-up entrepreneurs, have emerged as the new royalty. With basic services "problem-solved," the urgency to improve upon other aspects of the public sector decreases for these wealthy urban dwellers. As does their motivation to act as citizens. The serendipity and friction of public life—a powerful source of romance—threatens to become eclipsed by the monoculture of privately engineered technology solutions.

Even the most private sphere of our lives, the final refuge of privacy—our dreams—are under attack now, thanks to a mobile app called Shadow that lets users record their dreams after they experience them. The dreams are stored and analyzed, and then aggregated along with those of thousands of other users. The result might one day be a vast dream database that allows for an intimate sentiment analysis and for predictive models that project what the users, and ultimately humanity, will dream. "Shadow" is a fitting name for the start-up because gaining access to our dreams is like a dark cloud looming over our sovereignty. What if our dreams, our most private intimations, become the most public expressions of ourselves for which we are held accountable in perpetuity? What if our dreams are shared, traded, or even "hacked"? What if we began to brand or trademark them? The idea that our dreams are becoming commodities is the romantic's ultimate nightmare.

Shadow is a precursor of more tracking apps to come. Dreams are just one part of the universal literacy that the apostles of the Quantified Self want to establish. They demand that everything that is dreamed should be analyzed, that everything that is written should be read, and that everything that is said should be heard. In short: they are determined to eliminate doubt. Yet, as the writer Graham Greene told us, "Doubt is at the heart of the matter."[11] This sentiment echoes one of the most powerful ideas of the Romantic Movement: "negative capability," which the poet John Keats, in 1817, defined as man's capacity for "being in uncertainties, mysteries, doubts, without any irritable reaching after fact and reason."[12] Doubt is a key pillar of our morality. "Abolish all doubt, and what's left is not faith, but absolute, heartless conviction," as the writer Lesley Hazleton put it, and from which it is only a small step to "the arrogance of fundamentalism."[13]

Today's technology even promises to eliminate doubt about our future. The incessant collection of metadata is busy creating composites, generating an analysis of us that the individual mind can no longer fully grasp. Our story is now part of a bigger puzzle that operates on the power of aggregation and predictive analytics. This metadata is not only used by companies and advertisers eager to sell us things, it is also used by businesses making predictions about us and our social lives, consistent with our past behaviors.

For romantics, all of this is unacceptable. It is unacceptable to store and analyze our data, to construct social graphs and link our faces to ads, to pry into the "digital exhaust" of millions of people, and for citizenship to be adjudicated by engineers and software developers. It is unacceptable not only because it violates our rights as residents of a democracy, but also because the problem-solving impulse of these tech innovators dampens the mystery, messiness, and surprise of life—our vision of what *could* be.

Data driven decision making and design may suggest the most efficient, the most convenient solution. But we need more than efficiency and convenience the fastest route to delivering on a need or finding an answer—if we want to create more than just utilitarian businesses, products, and services. Data might be able to predict new problems or find new solutions to existing problems, but it will rarely imagine solutions to problems that do not yet exist. Only the human imagination—and its sense of wonder—can exceed the limits of mere solution. This excessiveness is what romance and innovation have in common. Both of them are a leap of faith—and a gift.

As Business Romantics, we celebrate the way data can help us tell new stories, learn more about ourselves, and lead us to possibilities and surprise. What we are against, however, is the maniacal belief in the doctrine of data. We romantics know it is only one of many truths. For that reason, we revolt when the world attempts to portray us through this one single narrative lens. We are so much more than our online track record, our social network influence, our activity "heat maps," and our "trust scores." We are human because we are unpredictable. We are human because we can't be trusted. Our inconsistency—our ability and desire to constantly change—is at the very heart of who we are. It is the only way to avoid becoming a data ideologue, to be a fan without turning into a fanatic. When everything is recorded, technology will steal "the romance of old conversations, that quaint notion that some things are best forgotten," as technology journalist Quentin Hardy points out.[14] He links the right to forget to the ability to start anew: "That quintessential American trait, self-reinvention, may well be threatened in the hard world of video and audio documentations and the chase of objective truth."

At a time when radical transparency and total datafication are the norm, we must reinvent ourselves as "romantic hackers." Like engineers use hacking to tweak products, services, or businesses in order to increase their value, Business Romantics can embrace hacking as an attempt to deepen the meaning of relationships. We can write romantic code that heals us from the uniformity of rationality: software that gives us erratic and emphatic experiences of "otherness." We can build, apply, or alter digital technology to expand the territory beyond our maps. We can create "adversarial designs"[15] that challenge our values and convictions, and we can launch services that disturb our concept of reality. We don't aim to make things more useful, only more beautiful. We don't want to answer all of our questions, solve all of our problems, or cure all of our pain. We just want to make them worthwhile.

And so it is at this point that we, as romantics, become public activists, occupying our own place between exuberant tech-optimism and purpose economy, radical self-expression and social good. We must create alternative avenues of meaning. We must subvert the suffocating surveillance of our age by utilizing all the tools in our romantic arsenal. With these tools, we can spearhead romantic campaigns at work and in our customer experiences, but also in the world at large. With our secret codes and small acts of significance, we can reclaim our narratives and character from the ceaseless scribes of Big Data. We can employ digital technology to rewrite the world around us or, at the very least, read it differently. We can bring friction back to seamless collaboration. We can restore doubt in a life of universal literacy and zero ambiguity, in a society aiming for total knowledge.

We have come full circle. I began this book with romance as a flame. At the end, romance is again lighting the way. In our work and customer experiences, it is casting shadows and creating depth. It is an invitation to see the hidden richness in our offices, conferences, trade shows, stores, and relationships. Whether you are a producer or consumer; an innovator or administrator; a leader, hermit, renegade, or foot soldier, ignite your flame and fight for your right to romance. You will not be alone. Let us view the world of business with fresh eyes. Let us start a fire in the dark, in the spaces between our busy comings and goings. Let us begin a new romantic age in which seeing, finding, and creating something greater than ourselves is everybody's business.

APPENDIX

The Business Romantic Starter Kit

Whether you are a cynic who is ready to convert, a romantic who wants to become a Business Romantic, a secret Business Romantic who is eager to come out, or a known and committed Business Romantic who wants to rekindle his or her spirit or recruit others, the following resources, tips, and activities will support you in getting started (or starting all over again). Romance, of course, defies prescriptive measures, so consider this Starter Kit a source of inspiration rather than a set of tools. Use it as a library that helps you connect not only with romantic ideas but also with other Business Romantics out in the professional field. You can find a constantly updated online version, along with the Business Romantic questionnaire, at www.thebusiness-romantic.com. My hope is that the Starter Kit will become an open-source platform by Business Romantics for Business Romantics.

The Business Romantic Playlist

SYLVAIN CHAVEAU *The Black Book of Capitalism*
Sylvain Chaveau is a French electronic music artist, and this 2000 work of his celebrates noir-esque meditations that draw from eclectic styles. The album's title is inspired by *Le Livre noir du capitalisme* (*The Black Book of Capitalism*), a book that was published in France in 1998 in reaction to *The Black Book of Communism* (1997) and featured critical essays on capitalism from various writers. Is it the perfect sound track for reflection on your professional life and on the capitalist system as a whole.

JONI MITCHELL *"A Case of You"*
With its enigmatic lyrics and lofty chords, this is perhaps the greatest love song ever written. "A Case of You" is pragmatic, bittersweet, even sarcastic, but still full of longing. The cover version by U.K. electroswooner James Blake is not bad either.

THE SMITHS *"Please, Please, Please Let Me Get What I Want This Time"*
Isn't this the very romantic reason we buy?

GUSTAV MAHLER *Symphony no. 5 in C-Sharp Minor, Adagietto*
The adagietto became famous through Luchino Visconti's movie *Death in Venice*, based on the novella by Thomas Mann, and it has lost nothing of its agonizing beauty. To tender harp chords, the melody seems to be wandering aimlessly and exposes the listener to a feeling of permanent suspension that is only resolved at the very end. The overall sentiment is one of otherworldliness, and the score instructions state, unmistakably: "Primary emotion: love."

ARMS AND SLEEPERS *Nostalgia for the Absolute*
The romantically titled album by this Maine-based duo features short one-to-three-minute electronic pieces inspired by or writ-

ten for movies: "Lovers Arctic," "Croix Rouge," "Clayton," "Crash," etc. Each track is like a fleeting moment of splendid isolation, a glimpse through a window to a world of wonders. "Beach music for the dead of night," one reviewer wrote.[1]

THE BLACK DOG *Music for Real Airports*

Referencing Brian Eno's legendary *Music for Airports* from 1978, a classic of the ambient music genre, British electronica trio Black Dog captures the sorrows of the businessman as a frequent flier with this 2010 album. It is the sound track for off- and on-hours during airport layovers, the quintessential physical and emotional no-man's-land. Tracks range from "Terminal EMA," "DISinformation Desk," "Passport Control," "Wait Behind This Line," to "Empty Seat Calculations," "Strip Light Hate," "Future Delay Thinking," "Lounge," "Delay 9," to "Sleep Deprivation 1" and "Sleep Deprivation 2," "He Knows," to, finally, "Business Car Park 9."

ARVO PÄRT *Cantus in Memory of Benjamin Britten*

This work by the Estonian composer Arvo Pärt is the manifestation of "infinite longing." One comment from a high school student on YouTube says it all: "Usually, no one really listens to the pieces of music our teacher presents to us. But this time, something was different. Everyone listened. No one spoke. And afterward, we couldn't speak for half an hour."[2]

VAN MORRISON *Astral Weeks*

This masterpiece by the Irish singer and songwriter evokes William Blake and other romantic poets, and features the famous lines "It ain't why, why, why; it just is." The highlight is "Beside You," a fervent ode to the other.

LEONARD COHEN *"Paper Thin Hotel"*

The Canadian bard describes in graphic detail the experience of hearing his wife make love to another man in the hotel room next door. It's a song about love, but also a song about listening. In the midst of the pain, Cohen creates a moment of self-revelation, of release and transcendence: "You go to heaven once you've been to hell."

ARNOLD SCHOENBERG *Transfigured Night*
Inspired by the poem of the same name by modernist German poet Richard Dehmel, this string sextet describes the dark secret a woman shares with her lover during a nightly walk: she carries the child of another man. Schoenberg's composition stretches the boundaries of late romanticism and transgresses into chromatic territory. The work was controversial not only because of some sexually explicit lyrics, but also because it featured a "nonexistent," unclassified chord, the mysterious "inverted ninth chord." Schoenberg commented: "And thus [the work] cannot be performed since one cannot perform that which does not exist."

T-BONE BURNETT *"Every Little Thing"*
T-Bone Burnett's raw acoustic tune is perhaps the most ascetic version of "suffering (a little)" in song format. Unsentimentally, it records the sweet-bitter aspects of having skin in the game: love has consequences.

JACQUES BREL *"Ne me quitte pas"*
This is the one song you give when you've given everything, the perfect endnote after everything's already over. There is no other love song that is so vulnerable, so sincere, and so completely free of irony.

THE ROMANTICS *"Talking in Your Sleep"*
There comes a time when keeping secrets keeps you up at night.

The Business Romantic Filmography

ECLIPSE *(directed by Michelangelo Antonioni)*
Set against the backdrop of the industrialist Rome of 1962, this seminal film by Italian director Michelangelo Antonioni is a poemlike meditation on the "modern malaise": the complications of finding love in the age of a rationalized and mechanized market society. It tells the story of a young woman and her affair with a stockbroker who is more fixated on material status than romantic pleasures. Their relationship ends one day, quietly, when neither of them shows up to a date.

LOST IN TRANSLATION *(directed by Sofia Coppola)*
Sofia Coppola's film translates Antonioni's themes of alienation and disconnection to the information age. A washed-up actor meets a young woman at the Park Hyatt hotel in Tokyo, and the two begin a relationship that is romantic without ever becoming physical. They are lost in translation, literally and metaphorically; in a hectic, über-consumerist society, this very limbo is the only space remaining for true emotions. As in every good love story, the end remains open: when the two "lovers" part ways, he whispers something into her ear that is eclipsed by the noise of urban Tokyo. To date, amid much speculation, even the best lip readers have not been able to detect what he told her.

MAD MEN *(created by Matthew Weiner)*
The award-winning TV series captures the glory days of the advertising business in the sixties on New York's Madison Avenue. Lead protagonist Don Draper is the quintessential romantic hero, a complex character and man of mystery who seeks conflict, thrill, and danger to heal an "old wound." Lies, betrayal, and abundant alcohol consumption become the col-

lateral damage of his all-consuming desire for love and identity. Then and now, work–life balance seems like an impossible idea, except that in the world of *Mad Men* the imbalance held the promise of an Old-Fashioned, of quick wins and profound losses, and other romantic adventures.

THE GODFATHER *(directed by Francis Ford Coppola)*
Is it cynical to consider *The Godfather* one of the most romantic movies of all time? Francis Ford Coppola's Mafia saga vividly contends that there is much more at stake in business than just business. In the end, it's a zero-sum game, with as many losers as winners; but material gain is never the only thing that matters to any of the players. The ambitions of today's "masters of the universe" on Wall Street and in Silicon Valley dwindle in the face of the Sicilians' absolute quest for family honor. It may not justify all means, obviously, but this intimate portrait of striving for something greater than oneself, for a legacy passed on from one generation to another, as well as the film's examination of the fragile joys of power, is a riveting parable about the fine line between romance and cynicism.

ROMAN HOLIDAY *(directed by William Wyler)*
This 1953 film is the mother of all romantic comedies and a congenial riff on almost all of the Rules of Enchantment. A foreign correspondent (Gregory Peck) meets an incognito princess (Audrey Hepburn) who has run away from the golden cage of her royal life. He spends a cheerful day with her in Rome, hoping to sell the biggest tabloid story of his life. Inevitably, they fall in love, and he eventually gives up on his plan, forgoing money for romance (and ultimately unfulfillment). The final scene, a press conference at the Palazzo Colonna, is unexpected and subtle: "Rome, by all means, Rome," the princess whispers in violation of diplomatic protocol when she's asked about

her favorite city on this trip. The movie denies both the couple and the viewers a fairy-tale happy ending—and the correspondent walks slowly out of the palazzo, all alone.

THE LAST TYCOON *(directed by Elia Kazan)*
Elia Kazan's 1976 movie, based on an unfinished novel by F. Scott Fitzgerald inspired by the life of legendary Hollywood producer Irving Thalberg, recounts the fall and rise of a larger-than-life movie mogul (Robert De Niro). His power erodes when he, the man in total control, falls in love with a young woman. The movie includes the famous scene in which De Niro schools a budding screenwriter with a theatrical background about "making pictures" by enacting a movie scene in his office. At the end, he walks alone into a dark and quiet production hall, like Gregory Peck walked out of the palazzo in *Roman Holiday*.

JIRO DREAMS OF SUSHI *(directed by David Gelb)*
Jiro One, the subject of this 2011 documentary, is considered by many to be the world's greatest sushi chef, and his restaurant, Sukiyabashi Jiro, a tiny ten-seater located in the tunnels of the Tokyo subway, was the first restaurant of its kind to be awarded the prestigious three stars by the Michelin guide. Jiro is remarkably meticulous and demanding, rarely taking a vacation and working from five in the morning until after midnight almost every day for the last seventy years. A former apprentice describes Jiro through the framework of the Japanese word *shokunin*. Though the word is directly translated into English as "artisan," in Japan, it implies something much deeper, as sculptor and woodworker Toshio Odate explains, "not only having technical skill, but also . . . an attitude and social consciousness. . . . The *shokunin* has a social obligation to work his or her best for the general welfare of the people. This obligation is both spiritual and material, in

that no matter what it is, the *shokunin*'s responsibility is to fulfill the requirement."[3]

CITIZEN KANE *(directed by Orson Welles)*
"Rosebud" is the last word the main character, newspaper mogul Charles Foster Kane (a character based on William Randolph Hearst) utters before his death. A journalist is put on the case, and he tries to find out "what it means," only to discover that he is facing an impossible task. Kane's sled, named Rosebud, was taken from him when he was separated from his family as a young child and sent to boarding school. Film critic Roger Ebert writes that *Citizen Kane* explains what Rosebud is, but not what it *means*.[4] He draws parallels to other iconic moments, "the green light at the end of Gatsby's pier; the leopard atop Kilimanjaro, seeking nobody knows what; the bone tossed into the air in *2001*; it is that yearning after transience that adults learn to suppress." "Maybe Rosebud was something he couldn't get, or something he lost," says the reporter assigned to the puzzle of Kane's dying word. "Anyway, it wouldn't have explained anything." Orson Welles's masterpiece is a remarkable illustration of our need for explanation and the poignancy of our constantly futile quest to achieve it.

FITZCARRALDO *(directed by Werner Herzog)*
Werner Herzog tells the story of the eccentric opera lover Brian Sweeney "Fitzcarraldo" Fitzgerald (played by the legendary German actor Klaus Kinski), who is determined to transport a steamship over a steep hill in the Amazon in order to access rich rubber territory. It is a tale of uncompromised passion, a character study of a romantic who gives everything and wins nothing.

HER *(directed by Spike Jonze)*
Her is a 2013 science-fiction fairy tale from the very near

future about wanting to find more romance in our hyper-connected lives, even if that means falling in love with an operating system. In one scene, "Samantha," the OS, reveals to her lover that she is simultaneously having conversations with 8,316 other users and that in fact she is in love with 641 of them. "I am yours, and I am not," she tells him, in not-so-subtle binary code. For her, this is not a problem: "The heart is not like a box that gets filled up; it expands in size the more you love. I'm different from you. This doesn't make me love you any less. It actually makes me love you even more."

The Business Romantic Reading List

F. SCOTT FITZGERALD *The Great Gatsby*
Fitzgerald's classic is a character study of a man who accumulates material wealth and social influence in a quest to find the love of his life. The novel pinpoints the allure and perils of success, the power of illusion, and the danger of losing oneself. Gatsby's enterprise is romance, and his means are those of a con artist. In the end, he fails, but his dream remains "incorruptible."

JULIO CORTÁZAR *Manuscrito hallado en un bolsillo*
This story by the legendary Argentinian writer is the mother of all "missed connection" stories. It recounts the fate of a man possessed by the memory of a woman he once spotted on a train. He spends the rest of his life chasing after her, haunted by a feeling of perpetual yearning.

PETER DRUCKER *The Essential Drucker*
Peter Drucker, who died in 2005, is widely considered the godfather of modern management theory; in fact, he might have

been the first to consider management an art. He developed the concept of purpose-driven business, embraced the importance of values, and, as the Renaissance man he was, insisted on linking management to the humanities. It might seem a little far-fetched to consider him a Business Romantic, but let's just say he is a kindred spirit and his work is as relevant now as it has ever been. When I studied business administration, reading Drucker soothed my soul; his voice sounded human and warm amid the cold logic of the corporation. He coined the term "knowledge worker" and proclaimed the "end of the economic man" long before others saw cracks in the veneer of neoclassical theory and profit-only business. Drucker heralded the possibility and responsibility of the modern corporation to build communities and create meaning. Many of his quotes have become common business wisdom: "Management is doing things right; leadership is doing the right things." . . . "If you want something new, you have to stop doing something old." . . . "The best way to predict your future is to create it." My favorite Drucker quote is lesser known: "So much of what we call management consists in making it difficult for people to work."

WALTER ISAACSON *Steve Jobs*

He is the obvious example, but it would be a grave omission to leave him out: Steve Jobs, the cofounder and former chairman and CEO of Apple, remains the archetypical Business Romantic—bold, relentless, risk-taking, always in search of something greater than himself. Walter Isaacson's biography portrays Jobs's bright and dark sides. Obsessing over details, Jobs knew that the interior of things, the soul of devices, mattered more than pure functionality, and his constant striving for beauty and perfection emanated from a desire to humanize technology and bring some magic, some romance, back to the launching of new products. A Howard Hughes–like man of mystery and a master at playing with concepts of presence and absence, Jobs used his "reality-distortion field" to deflect naysayers and pragmatists. Jobs was a fool who saw the world as it is *not* but *could be*—and he certainly lived up to his own promise: "Real artists ship."

ALAIN DE BOTTON *The Pleasures and Sorrows of Work*
The popular Swiss-British philosopher examines the modern workplace and its claim to serve as a principal source of meaning in ten reportage-style portraits of different professionals and professions, from an industrial biscuit bakery in Belgium to the soulless headquarters of one of the world's largest accounting firms. De Botton contends that a job is meaningful "whenever it allows us to generate delight or reduce suffering in others." But he knows how difficult this is to achieve: "It is surely significant that the adults who feature in children's books are rarely, if ever, Regional Sales Managers or Building Services Engineers."

JOHANN WOLFGANG VON GOETHE *The Sorrows of Young Werther*
I wish I had read it. And I hope you will. While it may be embarrassing to admit that I have not even held this book in my hands, I assure myself that there is something romantic about the prospect of reading it one day.

HERMAN MELVILLE *Moby-Dick*
The monomaniacal Captain Ahab becomes entirely absorbed by a single mission: the hunt for Moby-Dick, the "big white whale," to take revenge for the leg it once took from him. Abandoning all reason, Ahab slowly lets go of his humanity, the ability to relate to mankind and empathize with fellow beings. This Great American Novel depicts the dark side of the romantic: the danger of losing oneself in a quest to capture the transcendent: "It is not down on any map; true places never are."

ADAM PHILLIPS *Missing Out: In Praise of the Unlived Life*
This beautifully dense book makes one big argument—that we should spend our time living the life we have rather than aspiring to the one we desire to have—and then spends the remaining pages revisiting it without ever becoming boring. It is replete with quotes (one could also say sound bites) that are so sharp that they often don't require lengthy explanations or commentary. Think of it as a highly eloquent conversation with yourself. A trained psychoanalyst, Phillips observes that we "spend a great deal of our lived lives trying to find and give the reason that our other lives were not possible. And what was not possible all too easily becomes the story of our lives." Fittingly, the book contains chapters "On Frustration" and "On Not Getting It," examining the lack of (self)-actualization, comprehension, and fulfillment. In the age of consumer capitalism, Phillips writes, knowing ourselves means "simply knowing what we want to have." He leaves us with hope: our frustrations can be more meaningful touch points with the world than our desires.

MAX FRISCH *Homo Faber*
This book by Swiss author Max Frisch is the story of engineer Walter Faber, a man who "like every real man . . . lives in his work." He epitomizes the anthropological definition of the *homo faber*, the "working man," as opposed to the *homo ludens*, the "playing man," who thrives on entertainment, humor, and leisure. For Faber, only the tangible, calculable, quantifiable, and verifiable exists. Through a perfect storm of uncontrollable forces, however, he finds himself suddenly confronted with emotions that challenge his long-cultivated rational worldview. Faber meets a young woman, Sabeth, to whom he is strangely attracted, realizing too late that she's his daughter. Hanna, Sabeth's

mother, poignantly characterizes Faber's obsession with technology: "Technology [is] the knack of so arranging the world that we don't have to experience it." As the avatar of this worldview, Faber is opposed, and for the most part immune, to the idea of experience. He is even more averse to the loss of control, including life's most extreme submission, death. Faber is keenly aware of this himself: "Primitive peoples tried to annul death by portraying the human body—we do it by finding substitutes for the human body. Technology instead of mysticism!" Faber is the antiromantic who needs to have his heart broken first, by his own flesh and blood, before he can begin to truly appreciate the dimensions he had excluded from his life. Through the experience of intensity and pain, he, the metaphorically "blind man," learns to see.

LESZEK KOŁAKOWSKI *"In Praise of Inconsistency"*
In this 1992 essay, Polish philosopher Leszek Kołakowski heralds inconsistency as a basic human right and insists on the essentially unquantifiable nature of human beings. In fact, he believes that inconsistency is vital to the moral integrity of societies: "Total consistency is tantamount in practice to fanaticism, while inconsistency is the source of tolerance." Thus, he concludes: "Humanity has survived only thanks to inconsistency."

MARK EDMUNDSON *Why Teach? In Defense of a Real Education*
The American English professor and author is one of the most fervent defenders of the humanities and a keen observer of "digital natives." His argument for liberal arts is a romantic one: "The English major reads because, as rich as the one life he has may be, one life is not enough." Edmundson understands teaching as a calling in which the "souls" of students

are at stake, as an urgent endeavor to help them rethink who they are and who they might become. He believes that great teachers can "crack the shell of convention," shining a light on a "life's different prospects." Rather than "conversion," they pursue what Ralph Waldo Emerson called "aversion"—bucking conformity so as to discover possibility. In response to companies' expectations, many schools' first and foremost mission is to educate for "employability," but when teaching becomes training, and classes "incubators," then the "life-thickening" transformational experience that Edmundson wants from education will dissipate and schools may end up propagating a cybernetic lifestyle—becoming as much like a machine as possible.

DAVID FOSTER WALLACE *Commencement Speech, Kenyon College, 2005*

The late writer's now-legendary speech on the value and virtue of a liberal arts education, made popular through its later video version titled "This Is Water," is a passionate, lucid reminder of the painful beauty of an ordinary life—and the compassion it requires: "The really important kind of freedom involves attention and awareness and discipline, and being able truly to care about other people and to sacrifice for them over and over in myriad petty, unsexy ways every day." This has "almost nothing to do with knowledge," he claims, and "everything to do with simple awareness."

LESLIE JAMISON *In Defense of Saccharin(e)*

Leslie Jamison's essay dissects sentimentality as our "sweetest fear"; in fact her own fear of sentimentality—"the luxury of an emotion without paying for it" (Oscar Wilde)—as the lurking revelation that life without "saccharine" might not be as sweet as desired. She reveals the banality of kitsch as the horror of our own banality, our fear that we are not at all distinct from one another but rather feel the same universal feelings, like Pavlov's dogs. And yet, examining the world of artificial sweeteners, "honeymoons," lattes with "too much cream," affectionate names such as "sweetie" and "honey," trashy romances, and other tearjerkers, Jamison insists on "the faith that there is something profound in the single note of honey itself."

WILLIAM DERESIEWICZ *Commencement Speech, U.S. Military Academy at West Point, 2009*

The 2009 graduates from the elite military academy must have been surprised when they learned about the commencement speaker's topic: "solitude and leadership" are an unlikely pair, especially in the context of military leadership. Deresiewicz takes his audience on a journey into the "heart of darkness" and explains why the ability to think alone is crucial for any leader. He points to a "crisis of leadership" and bemoans it as a crisis of thinking, in fact, a lack of thinkers: "People who can think for themselves. People who can formulate a new direction: for the country, for a corporation or a college, for the Army—a new way of doing things, a new way of looking at things. People, in other words, with *vision*." This vision doesn't come from constant immersion in information. It comes from true reflection, Deresiewicz argues, from slowing down. A thinker is someone for whom thinking is more difficult than for other people, he claims, paraphrasing a line from German novelist Thomas Mann, who said something analogous about writers.

GEORGE SAUNDERS *Convocation Speech, Syracuse University, 2013*

In this both heartfelt and witty speech, writer George Saunders discusses the concept of "success" and admits that what he regrets most in his life are "failures of kindness." The reason is simple: "Who, in your life, do you remember most fondly, with the most undeniable feelings of warmth? Those who were kindest to you, I bet." For Saunders, the idea of a good life is synonymous with an expansive form of kindness: some may also call it love. Most people, he claims, become less selfish and more loving as they age, ultimately becoming "replaced by LOVE." He cites the poet Hayden Carruth, who, in a poem near the end of his life, wrote that he was "mostly Love, now."

The Business Romantic Destinations

BURNING MAN *Black Rock Desert, Nevada*
Every year, more than fifty thousand people—also called "Burners"—embark on the pilgrimage to the Black Rock Desert in Nevada for a week-long experiment in radical self-expression, self-reliance, and art. Burning Man is the most genuine manifestation of the gift economy, and a Dionysian act of transformation. It reaches its climax with the ritual burning of a large wooden effigy. The Burning Man's ten principles include giving, de-commodification, and immediacy; the Business Romantic finds common cause in almost every one of them.

MLOVE *South of Berlin, Germany, and Monterey, California*
Founded by German serial entrepreneur Harald Neidhardt, the MLove ConFestival, with annual events at a castle south of Berlin and in Monterey, provides an eclectic community of curious business minds from the tech, media, and entertainment industries with an intimate forum for genuine exchange. Now in its sixth year, the festival has gradually shifted

its focus from "M-obile" to "M-eaning," and speakers and attendees consistently manage to bridge the gap between practical business issues and big philosophical questions. MLove is a mix of campfire, learning expedition, wedding weekend, and business conference: romantic and pragmatic at the same time—the quintessential Business Romantic event.

HOW THE LIGHT GETS IN *Hay-on-Wye, England*
The world's largest philosophy and music festival takes place in the "town of books": Hay-on-Wye in England. Philosophers, writers, poets, artists, and a few scattered business leaders discuss topics such as "Why Is There Something Rather Than Nothing?" "In Search of Lost Time," and "The Limits of Logic" during the day, and then dance their big ideas and deep thoughts away with fun parties at night. In 2014, a panel debated the pros and cons of the romantic ideal of love: "Might we live more fulfilling lives if we gave up chasing this romantic ideal or does it still offer us the most exciting adventure of our lives?" Hard to imagine that anyone at this gathering would side with the former.

GRIFFITH PARK OBSERVATORY *Hollywood*
A hike above the busyness of Hollywood, this recently reopened observatory became iconic through Nicholas Ray's classic movie *Rebel Without a Cause* and offers one of the most stunning panoramic views of L.A., from downtown to the Pacific Ocean. It is no easy feat to find serenity amid L.A.'s urban mall, but standing on the deck up here makes you rise above the wheeling-and-dealing. If you can make it up here, you can make it everywhere.

CAMP NOU STADIUM *Barcelona*
Watching a soccer game at this one-hundred-thousand-seat stadium, the home grounds of FC Barcelona, or Barça, is like going to the opera. Before each game, the Barça anthem

is intoned by the crowd, and seventeen minutes and fourteen seconds into the game, the fans begin to cheer "independence!," acknowledging the end of Catalan independence in 1714. In contrast to other soccer cathedrals, though, the stadium is generally not very loud, and a certain detachment is palpable in the stands. It is not amusement, it is art; and the connoisseurs occasionally applaud an outstandingly elegant pass or movement, but for the most part just sit back and enjoy the unfolding spectacle. Like most romantics, the Barça players work hard, but their effort never looks industrious.

The Business Romantic Action Items

COME OUT AS A BUSINESS ROMANTIC

If you feel that the workplace or a brand does not accommodate your full range of spiritual, emotional, and intellectual needs, it is time to come out as a Business Romantic. It is imperative that you don't just "confess" (for example, don't open the conversation by stating that you think your workplace or a brand may not be "romantic enough"; also refrain from inviting your boss to a candlelight dinner). You can start by sending subtle signals: play your favorite music (or tunes from the Business Romantic playlist) through your office speakers. Surprise your colleagues and friends with a reference to a romantic writer or movie. Tell them about a moment at work or a customer experience, seemingly small and fleeting, that meant a lot to you. Oh, and of course, you can simply leave a copy of this book on someone's desk. One day the recipient will approach you with a smile of recognition: "I knew it!"

SPOT OTHER BUSINESS ROMANTICS

Look for the subtle, often nonverbal cues. An expressed interest in the (liberal) arts is usually a good sign. Generosity and a passion beyond work, too. You might identify an aura of secrecy, unreasonable behavior, excessive attention to detail, and other aspects of strangeness. When you're sure you have found a fellow Business Romantic, say nothing. Everything will fall into place.

GO ON A BUSINESS DATE WITH ANOTHER BUSINESS ROMANTIC

Once you have come out as a Business Romantic, and you have identified other Business Romantics within or outside your organization, this one should be easy. You have a lot to talk about. Just make sure you don't overshare and keep up the mystique. Broken hearts stay in the closet. No one wants to hear about past business relationships or flings. Trust may be built through a portfolio; attachment isn't. It's not about what you did but what you want to do together.

GO ON A BUSINESS DATE WITH A BUSINESS CYNIC

This one is harder because many cynics are secret romantics. So double-check and do some extra due diligence before you reach out. Once you are sure that your date is a true cynic, the fun begins. At dinner, warm him or her up with personal anecdotes, then shock your date with some poetry before entrusting them with a secret you have not shared with anyone else. That will break the ice. Ask your date about the moment in which they felt the most alive. Ask them what they are afraid of and why. And then just listen.

PRETEND TO BE A BUSINESS ROMANTIC

Want to be a superhero for a day? Be a Business Romantic! Step into the shoes and psyche of your alter ego and embody and enact the Business Romantic's Rules of Enchantment, just for fun, just for a week. Or even just for a day. Be a hermit, a rebel, a contrarian, a poet. Be a stranger to yourself. Masks transform us, and yes, you can fake it until you become it!

START A SECRET PROJECT (WITHOUT REASON)

Call a meeting, preferably off-site, and invite a small group of colleagues. Don't tell them the reason for the meeting, and even when you meet, do not pretend there is one. Just tell them that you would like to collaborate with them on a project that would "add value to the company and move the business forward." And then spend the whole kickoff meeting exploring together what that might be. It's a bit like Luigi Pirandello's play *Six Characters in Search of an Author,* except that here you are in search of a project. Easy!

USE THE BUSINESS ROMANTIC JOB POSTING

Gradually incorporate elements of the Business Romantic description into each of your new job postings; use the Business Romantic job posting to hire for any position; ultimately, hire a full-time Business Romantic. And then a full Business Romantic team!

HOST A BUSINESS ROMANTIC DINNER

Dinner is the most congenial and appropriate way to connect with other Business Romantics. It doesn't have to be a full-blown candlelight dinner, but a private setting in a dimly lit and quiet environment certainly helps. Limit it to fifteen guests or less. It is very difficult if not impossible to have intimate conversations—and that is what you want—with more than fifteen people around one table. Invite a diverse mix of quests, by background and personality (for example, you don't want only "conversation commanders," but you certainly need some to get the flow going). Send out printed invitations (they make everybody feel special and communicate that a no-show will be paid for with lifelong guilt), and pick a topic that relates directly or indirectly to Business Romance. Create a seating arrangement. Part of your role as a host is to make sure that your guests are in good company. They trust you to find the right neighbor for them, so don't shy away from that responsibility. Pay attention to everyone. Make your guests smile, no matter how serious the topic. Have fun!

HOST BUSINESS ROMANTIC LUNCHEONS WITH YOUR COWORKERS, PARTNERS, OR CUSTOMERS

Plan small lunches around a Business Romantic topic and invite a small group of colleagues, partners, or customers. Keep it personal and ask specific questions, for example: "What was the most romantic moment in your life and why? What can our company learn from it?" Discuss what it means to have a romantic relationship to work or to a brand. Follow up with a handwritten note. Do it again.

TAKE ON THE BIG QUESTIONS

Do we want to live in a Business Romantic society? How is it different from a romantic society? How big and how inclusive do we want this society to be? How much romance can a prosperous society handle? How do you design (for) a romantic society and its communities (homes, campuses, cities, and workplaces)? What are the policies that support and inform a more romantic citizenry? Or will policies stifle it? Can you imagine romantic law and government? What does the Business Romantic city look like? Like the stereotypical romantic city—Rome, Paris?—or more like Dakar, Bogota, Detroit, or Columbus, Ohio? Is there romance—Business Romance?—possible in the suburbs, in the small towns, and in the countryside? These and other questions are the perfect topics for Business Romantic lunches and dinners or other related events.

KEEP A DIARY OF THE BRANDS THAT MAKE YOUR DAY

From the smile of a Starbucks barista in the morning to the supersmooth check-in on a flight with Lufthansa to the Volkswagen commercial that makes you laugh in the living room at night: honor the products and services that leave you wanting more.

KEEP TRACK OF WHAT YOU FAIL TO ACHIEVE EVERY YEAR

Announce it at your firm's holiday party and celebrate it with a toast.

SMUGGLE A ROMANTIC ELEMENT INTO A STANDARD PIECE OF COLLATERAL

It can be a visual motif, a line from a poem, or simply language that opens up space for ambiguity and mystery when you least expect it. Annual reports are suitable publications, and so are feature sheets and sales brochures.

MAKE A LIST OF YOUR SECRET LOVES

Keep it in a safe place. Don't say a word. Ever. To. Anyone.

INVITE A PERSON YOU WANT TO DO BUSINESS WITH TO MEET WITH YOU ON APRIL 1 EVERY YEAR (AT AN UNDISCLOSED LOCATION)

This is a classic romantic format, repopularized by movies such as *Before Sunset* or *One Day* (based on the bestselling novel). The April 1 date raises the stakes. Will they really show up (again and again)?

HOST THE BUSINESS ROMANTIC AWARDS

Invite your colleagues to a ceremony at which you award the greatest Business Romantics in your firm—once a year, every quarter, or perhaps even every month.

ENROLL IN THE BUSINESS ROMANTIC SCHOOL—OR START ONE!

A university for students for whom learning is (self-)discovery rather than self-advancement, the Business Romantic School teaches a romantic approach to business and cultivates inquiry and rebellion instead of ambition and conformity. It confronts you with the very low and the very high, the very private and the very public aspects of doing business. Carefully curated experiences of friction, online and offline, replace frictionless learning. The school does not have a permanent location, no dean, and only guest faculty. Students are teachers, and teachers are students. Each share stories from their career, seminal moments, romantic experiences, tips and tricks, nuts and bolts, do's and don'ts, victories and defeats, moments of vulnerability, expression, friendship, and love. "Classes" may take place around a dinner table, on excursions, or on "blind dates" with other romantics.

The school teaches you about the benefits of opacity, mystery, and secrecy, and how to build a fulfilling career on the merits of permanent unfulfillment. One night you will see an Antonioni movie that "explores the conditions of modern life in the contours of business," another night you will learn how to end projects, employment, and relationships with grace and humility. You will prepare for the first and last day on the job. You will practice generosity and devote time to the kindness of strangers. You will learn to recognize the hidden costs of unromantic business and to appreciate the hidden treasures of romantic business. You will spend more time on crafting a secret plan B than an obvious plan A, and you will learn how to enjoy moments of meaning and beauty without the need to monetize them. You will relearn how to write handwritten letters and sincere memos, and how to read between the lines of corporate lingo. You will conduct customer research and exhibit the results in museums dedicated to their subjects. You will learn how to be unreasonable and reasonably foolish. You will find out about your blind spots and how to fall in and out of love because of them. You will practice the art of doing nothing by simply sitting still. You will learn all the things that you do not learn at traditional business school but are critical for having a richer, more meaningful, more romantic life in business; all the things that are not listed on your résumé but offer you the possibility of having a second, richer life in business.

JOIN THE BUSINESS ROMANTIC SOCIETY

The Business Romantic Society is not a formal organization—it is a secret network of like-minded individuals worldwide who meet on an ongoing basis, gather at Business Romantic dinners, exchange "love stories," and discuss best practices. There will be an annual excursion. Still, you need to be invited. I'm at tim@thebusinessromantic.com. Send me an e-mail.

ACKNOWLEDGMENTS

My favorite experience in the movie theater is that last moment when the screen turns dark and the closing credits roll to the sounds of the final tune. With eyes still beaming, the lights still out, a silenced audience grants itself the permission to speak again, slowly realizing what it has seen. I often decide then whether I liked the movie or not. When everything is said and done, it is easier to romanticize.

So here are my closing credits, the moment I've been waiting for. As I was writing them, it occurred to me that it might serve us well—as workers, consumers, and citizens—to begin each project, each tenure, each life endeavor with a draft of acknowledgments, pondering the question "Whom, in the end, would you like to thank and why?" rather than "What would you like to have accomplished?" This will hone our humility, our ability to estimate our own position in the world more realistically. It will also bring to the forefront the social capital generated by any collaborative work effort as the ultimate return-on-investment, as an explorative, not exploitative enterprise: the kind of "sympathetic exchange" that philosopher Robert C. Solomon talked about. That's really what this book—any book—has been and is all about: a romantic enterprise in and of itself, the living proof that true romance is possible within the conditions of the market society.

It feels good to know you're not alone. I guess it has become a cliché to state that it has become a cliché to state that a book is a team effort. But it's true. It's one of the many things this project has taught and brought to life for me. *The Business Romantic* was a fling at first, a Business Romantic start-up, brittle, fickle, and, like all romantic endeavors, extremely vulnerable, but with every new collaborator, it gradually evolved into something bigger: The Business Romantic Society, a not-so-secret tribe, a collective of kindred spirits.

It began with an essay I wrote for a magazine back in 2009 on the new "meaning economy" and companies as "arbiters of meaning." But it took me a few more years and the help of Priya Parker to understand that this topic was more than a passion project: it was my cause worth fighting for. I met Priya at a dinner in New York in 2012, and through the two Visioning Labs she ran with me that summer, the idea for the book solidified.

Priya also designed and facilitated an Inner Circle Lab in Munich in January 2013 in which I presented my initial concept for the book to a small group of fellow writers and friends, unofficially and quite pretentiously dubbed "The Meaning Crew": Markus Albers, Gianfranco Chicco, Alexa Clay, Maggie De Pree, Malvina Goldfeld, John Havens, Andrian Kreye, Alan Moore, Ximo Peris, Navi Radjou, and Sagarika Sundaram. Thank you for being there from start to finish, for all your amazing help and support, honest criticism, and ideas. Love you, guys! And thank you, Priya, for holding the space for us, for your mentorship and friendship.

Furthermore, I would like to thank:

Seth Matlins—for CMO b-romance and the generous introductions to a secret network of Business Romantics.

Ernest Beck—for tough editing love.

Bruno Giussani, Liba Rubenstein, Angie Lee, Meghan O'Rourke, Jerri Chou, Chris Muscarella, June Cohen, Mark Barden, Kevin McSpadden, Tex Drieschner, David Naylor, Joseph Newfield, Liz Kelly, Emily Chong, Adam Richardson, Evan Selinger, Laura Gamse, Kal Patel, Liz Maw, Brian Behlendorf, Axelle Tessandier, Erica Williams, Niels Harper, Amy Lazarus, Matt Lerner, Harald Neidhardt, Rimjhim Dey, Christian Madsbjerg, Laura Galloway, George Bennett, Elyssa Dole, Kevin O'Malley, Christie Dames, Anand Ghiridharadas, Ansgar Oberholz, Nora Abousteit, Morgan Spurlock, and Dev Patnaik: for pointing me to the right people, sharing your stories with me, for the inspiring conversations, and for editing my raw ideas.

Gianpiero Petriglieri and David Kim for making me dig deeper.

My fellow members of the WEF Values Council for shaping my thinking and the many debates about business and its role in society, in particular Jim Wallis, Stewart Wallis, Michael Gerson, Daniel Malan, Michèle Mischler, and Daniel Shapiro.

My friends in Germany: Julian de Grahl, Niki Hauser, Benjamin Schlez (one day we'll do our own song collection), Matthias Braun, Jakob Hesler, Lars Precht, Saskia Rettig, Sascha Seifert, Nicole Ackermann, Carmen "Citi" Stephan, Fredy Osterberger, Martin Zünkeler, and Stephan Trüby for playing with me, listening, and giving advice when it mattered.

Wolfram Knöringer for being a fan, of this project and life in general.

Till Grusche for all the encouragement and support; and the many great moments at Frog Design, Martha's, on the conference floor, at TED Salons and in dive bars, in the soccer stadium, on the tennis court, and beyond.

Doreen Lorenzo for being my longtime fearless leader and for allowing me to embark on this project in the first place.

My fellow former frogs: Andrea Bebber, Jan Chipchase, Ravi

Chhatpar, Sally Dang, Robert Fabricant, Jaleen Francois, Mark Gauger, Eric Hummel, Cyrus Ikpachi, Reena Jana, Kristina Loring, Marie Lozano, Sam Martin, Sara Munday, Chloe Ng, Mark Rolston, Fabio Sergio, Kate Swann, Christian Schluender, and the many others who helped me swim in the pond.

Niklas Hofmann for the über-excellent German translation, and Melanie von Marschalck, Nadine Oberhuber, Jana Gioia Baurmann, Julia Mooney, and Daniel Haas for being part of this journey.

Darius Ramazani for early-years pictures.

Beowulf Sheehan for romantic photography and a perfect day in San Francisco.

Megan Lynch for art direction and design on all channels.

Brian Harris for all web development.

Mark Fortier and Norbert Beatty at Fortier PR for spreading the word.

Zoe Bohm, my editor at Little, Brown in the U.K., as well as Sarah Shea and Clara Diaz in marketing/PR; Margit Ketterle and Thomas Tilcher at Droemer Knaur in Germany, and their marketing/PR team: Esther von Bruchhausen, Harriet von Stauffenberg, and Christina Schneider; Petra Eggers, my literary agent in Germany; and Robert Kirby from United Agents in the U. K., for their enthusiasm and support.

Everyone at NBBJ, in particular the fabulous marketing team and Kay Compton, Jennifer Chobo, Rich Dallam, Helen Dimoff, Jay Halleran, Michael Kreis, Steve McConnell, Ryan Mullenix, Meghan Novak, John Pangrazio, Thurston Roach, Joan Saba, Tom Sieniewicz, Mackenzie Skene, Sally Suh, Jonathan Ward, and Scott Wyatt, for their collegiality, support, and inspiration.

Sarah Levitt at the Zoë Pagnamenta Agency for emails that always made my day. And a huge thank you to Zoë Pagnamenta for giving me a chance and being such an honest adviser throughout the entire process. I could not have hoped for a more dedicated literary agent.

The same holds true for Hollis Heimbouch, my publisher and editor at HarperCollins: Dear Hollis, thank you so much for seeing the book the way you did (even before there was much to see) and being with me, quite literally, line by line, from start to finish. It is a true delight to be working with you.

I'm also very grateful to the team at HarperCollins: Eric Meyers (associate editor), Lydia Weaver (production editor), Stephanie Cooper (marketing manager), William Ruoto (interior page designer), Robin Bilardello (cover design), Steven Boriak (publicity director). Your attention and care was terrific. Big hearts show themselves in the smallest details.

Ann Marie Healy, my editor and writing partner-in-crime: Boy, was I lucky. You could do magic and pulled a rabbit out of a hat when I needed it (and I needed it often). You whipped things into shape, always unfazed, steady. Thank you for being there through many false starts and re-starts, beginnings and endings, and for making the middle so much fun. Thank you for sailing the ocean together.

My parents-in-law, Charlie and Blair Moser, for welcoming me into their family. Blair, for being the toughest proofreader one could wish for (although I was surprised that your comments sounded way more like Milton Friedman than I would have suspected from a San Francisco–based writer).

Frank Leberecht, for editing rigorously, keeping me honest, and calling my bluff from time to time (just as you would expect from a little brother).

My mother, Edith. I miss you. My father, Volker, for carrying on the flame.

Harper Ava, my daughter and my Lieblingsmädchen.

And thank you, my love, my wife, and my best friend Sarah "Sarah Moser" Moser: you will always be the greatest romance of all.

San Francisco, June 5, 2014

NOTES

INTRODUCTION: THE FLAME

1. Max Delbruck, *Mind from Matter* (Oxford: Blackwell Publishers, 1985).
2. David Brooks, "The Olympic Contradiction," *New York Times*, July 26, 2012, http://www.nytimes.com/2012/07/27/opinion/brooks-the-olympic-contradiction.html.
3. Robert C. Solomon, *A Better Way to Think About Business: How Personal Integrity Leads to Corporate Success* (Oxford: Oxford University Press, 1999).
4. Alia McKee and Tim Walker, "State of Friendship in America," http://getlifeboat.com/goodies/report2013/.
5. David Whyte, *Crossing the Unknown Sea: Work as a Pilgrimage of Identity* (New York: Riverhead Books, 2001).

CHAPTER 1: THE NEW DESIRE FOR ROMANCE

1. Gallup, "State of the Global Workplace" (Gallup, 2013), http://www.gallup.com/strategicconsulting/164735/state-global-workplace.aspx.
2. Edelman, "2013 Edelman Trust Barometer Executive Summary" (Edelman, 2013), Scribd edition.
3. "World Economic Forum Top 10 Trends of 2014," World Economic Forum, 2013, http://reports.weforum.org/outlook-14/view/top-ten-trends-category-page/.
4. Stefan Steinberg, "OECD Reports Growing Inequality," World Socialist Website, May 17, 2013, http://www.wsws.org/en/articles/2013/05/17/oecd-m17.html.
5. Thomas Piketty, *Capital in the 21st Century* (Cambridge: Belknap Press, 2014); by the time of writing this book, the *Financial Times* had accused Piketty of deriving his findings from inaccurate data, with a fierce public debate ensuing: Chris Giles, "Piketty findings undercut by errors,"

Financial Times, May 23, 2014, http://www.ft.com/cms/s/2/e1f343ca-e281-11e3-89fd-00144feabdc0.html#axzz32eqg7ENQ.

6. Emmanuel Saez, Gabriel Zucman, "The Distribution of US Wealth, Capital Income and Returns since 1913," http://gabriel-zucman.eu/files/SaezZucman2014Slides.pdf.
7. David Bollier, "Power Curve Society: The Future of Innovation, Opportunity and Social Equity in the Emerging Networked Economy" (Washington, D.C.: the Aspen Institute, 2013), http://www.aspeninstitute.org/policy-work/communications-society/power-curve-society-future-innovation-opportunity-social-equity?utm_source=as.pn&utm_medium=urlshortener.
8. David Brooks, "Capitalism for the Masses," *New York Times*, February 20, 2014, http://www.nytimes.com/2014/02/21/opinion/brooks-capitalism-for-the-masses.html.
9. Jaron Lanier, *Who Owns the Future?* (New York: Simon & Schuster, 2013).
10. Marc Andreessen, "Why Software Is Eating the World," *Wall Street Journal*, August 20, 2011, http://online.wsj.com/news/articles/SB10001424053111903480904576512250915629460.
11. Nelson D. Schwartz, "The Middle Class Is Steadily Eroding. Just Ask the Business World," *New York Times*, February 2, 2014, http://www.nytimes.com/2014/02/03/business/the-middle-class-is-steadily-eroding-just-ask-the-business-world.html?_r=0.
12. Barack Obama, "Remarks by the President on Economic Mobility," December 4, 2013, http://www.whitehouse.gov/the-press-office/2013/12/04/remarks-president-economic-mobility.
13. Greg Smith, "Why I Am Leaving Goldman Sachs," *New York Times*, March 14, 2012, http://www.nytimes.com/2012/03/14/opinion/why-i-am-leaving-goldman-sachs.html?pagewanted=all.
14. Sam Polk, "For the Love of Money," *New York Times*, January 18, 2014, http://www.nytimes.com/2014/01/19/opinion/sunday/for-the-love-of-money.html.
15. "Work, Stress, and Health," National Institute for Occupational Safety & Health Conference, 1999.
16. Pew Research Center, "Millennials in Adulthood: Detached from Institutions, Networked with Friends," March 7, 2014, http://www.pewsocialtrends.org/files/2014/03/2014-03-07_generations-report-version-for-web.pdf.
17. Evgeny Morozov, "The Perils of Perfection," *New York Times*, March 2, 2013, http://www.nytimes.com/2013/03/03/opinion/sunday/the-perils-of-perfection.html?pagewanted=all; also see his book *To Save Everything, Click Here: The Folly of Technological Solutionism* (New York: Public Affairs, 2013).
18. Naomi O'Leary, "Pope Attacks 'Tyranny' of Markets in Manifesto for

Papacy," Reuters, November 26, 2013, http://www.reuters.com /article/2013/11/26/us-pope-document-idUSBRE9AP0EQ20131126.

19. Brigid Schulte, *Overwhelmed: Work, Love, And Play When No One Has The Time* (New York: Sarah Crichton Books/Farrar, Straus and Giroux, 2014).
20. Alia McKee and Tim Walker, "State of Friendship in America," http: //getlifeboat.com/goodies/report2013/.
21. Pew Research Center, "Millennials in Adulthood: Detached from Institutions, Networked with Friends," March 7, 2014, http://www. pewsocialtrends.org/files/2014/03/2014-03-07_generations-report-version-for-web.pdf.
22. Jeffrey Sachs, "The New Progressivism," *New York Times*, November 12, 2011, http://www.nytimes.com/2011/11/13/opinion/sunday/the-new-progressive-movement.html?_r=0.
23. "B-Corps," http://www.bcorporation.net, accessed March 2, 2014.
24. "Changemakers," http://www.changemakers.com/intrapreneurs, accessed March 2, 2014.
25. Deloitte University Press, "A Movement in the Making," January 2014, http://dupress.com/articles/a-movement-in-the-making/.
26. "The B Team," http://bteam.org/about/vision/, accessed February 25, 2014.
27. "Conscious Capitalism," http://www.consciouscapitalism.org/, accessed May 23, 2014.
28. "The Energy Project," http://theenergyproject.com, accessed May 29, 2014.
29. Aaron Hurst, *The Purpose Economy: How Your Desire for Impact, Personal Growth, and Community Is Changing the World* (New York: Russell Media, 2014).
30. Peter Drucker, *The Essential Drucker: The Best of Sixty Years of Peter Drucker's Essential Writings on Management*, Collins Business Essentials (New York: HarperBusiness, 2008).
31. Megan Webb-Morgan, "The New Work Style of Generation Y," *Society for Human Resource Management*, October 15, 2012, http://www. weknownext.com/workforce/the-new-work-style-of-generation-y.
32. Elizabeth Dunn and Michael Norton, *Happy Money: The Science of Smarter Spending* (New York: Simon & Schuster, 2013).
33. Tony Schwartz, Christine Porath, "Why You Hate Work," *New York Times*, May 30, 2014, http://www.nytimes.com/2014/06/01/opinion/ sunday/why-you-hate-work.html.
34. Deloitte, "Deloitte Survey: Strong Sense of Purpose Key Driver of Business Investment," press release, April 7, 2014, http: //www.deloitte.com/view/en_US/us/press/Press-Releases /f2ca7e803a725410VgnVCM1000003256f70aRCRD.htm.
35. United Nations Regional Information Centre for Western Europe, http://www.unric.org/en/happiness/27709-the-un-and-happiness.

36. Justin Fox, "The Economics of Well-Being," *Harvard Business Review*, January/February 2012, http://hbr.org/2012/01/the-economics-of-well-being/ar/1.
37. "Gross National Happiness," http://www.grossnationalhappiness.com, accessed February 26, 2014.
38. "Delivering Happiness," http://www.deliveringhappiness.com, accessed March 2, 2014.
39. Emily Esfahani Smith and Jennifer L. Aaker, "Millennial Searchers," *New York Times*, November 30, 2013, http://www.nytimes.com/2013/12/01/opinion/sunday/millennial-searchers.html?pagewanted=1&utm_medium=App.net&_r=0&partner=rss&emc=rss&utm_source=PourOver.
40. Arianna Huffington, *Thrive: The Third Metric to Redefining Success and Creating a Life of Well-Being, Wisdom, and Wonder* (New York: Harmony, 2014).
41. Viktor E. Frankl, *Man's Search for Meaning* (Boston: Beacon Press, 1959).
42. Ariel Schwartz, "Millennials Genuinely Think They Can Change the World and Their Communities," *Fast Company*, June 27, 2013, http://www.fastcoexist.com/1682348/millennials-genuinely-think-they-can-change-the-world-and-their-communities.
43. "Adhocracy," P2P Foundation, http://p2pfoundation.net/Adhocracy, accessed February 25, 2014.
44. Max Weber, *The Protestant Ethic and the Spirit of Capitalism*, translated by Peter Baehr and Gordon C. Wells based on original edition from 1905 (New York: Penguin Books, 2002).
45. John Gantz and David Reinsel, "The Digital Universe in 2020: Big Data, Bigger Digital Shadows, and Biggest Growth in the Far East," *IDC View*, December 2012, http://idcdocserv.com/1414. Accessed March 26, 2014.
46. Pico Iyer, "The Folly of Thinking We Know. The Painful Hunt for Malaysia Airlines Flight 370," *New York Times*, March 20, 2014, http://www.nytimes.com/2014/03/21/opinion/the-painful-hunt-for-malaysian-airlines-370.html.
47. Alex Pentland, *Social Physics: How Good Ideas Spread—The Lessons from a New Science* (New York: Penguin Press, 2014).
48. Ned Resnikoff, "Glassholes at Work," *The Baffler*, May 14, 2014, http://www.thebaffler.com/blog/2014/05/glassholes_at_work.
49. F. Scott Fitzgerald, "The Crack-Up," *Esquire*, February 26, 2008 (originally appeared 1936), http://www.esquire.com/features/the-crack-up.
50. "The Two Cultures," Wikipedia, http://en.wikipedia.org/wiki/The_Two_Cultures, accessed February 25, 2014.
51. Leon Wieseltier, "Perhaps Culture Is Now the Counterculture," *New*

Republic, May 28, 2013, http://www.newrepublic.com/article/113299/leon-wieseltier-commencement-speech-brandeis-university-2013.

52. "The Teaching of the Arts and Humanities at Harvard College: Mapping the Future," Harvard University, May 31, 2013, http://artsandhumanities.fas.harvard.edu/files/humanities/files/mapping_the_future_31_may_2013.pdf, accessed March 17, 2014.
53. David Silbey, "A Crisis in the Humanities?," *The Chronicle of Higher Education*, June 10, 2013, http://chronicle.com/blognetwork/edgeofthewest/2013/06/10/the-humanities-crisis/. Silbey suggests that Harvard's numbers are flawed because they don't include degrees in history or visual and performing arts, among other subjects. He also takes issue with the asserted ratio to "total degrees" in the United States, arguing that "we give out far more population-normalized degrees in the humanities now than we did in the 1950s or the 1980s." Lastly, Silbey points out that the 1960s experienced an extraordinary and unprecedented spike in humanities degrees and therefore don't serve as a valid baseline.
54. John van Doren, "The Beginnings of the Great Books Movement at Columbia," Living Legacies: Great Moments and Leading Figures in the History of Columbia University, *Columbia Magazine*, Winter 2001, http://www.columbia.edu/cu/alumni/Magazine/Winter2001/greatBooks.html.
55. Erik Brynjolfsson and Andrew McAfee, *The Second Machine Age: Work, Progress, and Prosperity in a Time of Brilliant Technologies* (New York: W. W. Norton & Company, 2014).
56. Tim Laseter, "Management in the Second Machine Age," *Strategy + Business*, Summer 2014, Issue 75, http://www.strategy-business.com/article/00252?pg=all.
57. Lee Foster, "1816—The Year Without Summer," National Weather Service, http://www.erh.noaa.gov/car/Newsletter/htm_format_articles/climate_corner/yearwithoutsummer_lf.htm, accessed February 26, 2014.
58. Mary Shelley, *History of a Six Weeks' Tour Through a Part of France, Switzerland, Germany, and Holland; with Letters Descriptive of a Sail Round the Lake of Geneva and of the Glaciers of Chamouni* (London: T. Hookham, Jr. and C. and J. Ollier, 1817), Google Books, accessed March 2, 2014.
59. Ibid.
60. Seán Manchester, "A Byronic Legacy," Gothic Press, http://www.gothicpress.freeserve.co.uk/Byronic%20Legacy.htm, accessed March 20, 2014.
61. Terry Castle, "Mad, Bad and Dangerous to Know," *New York Times*, April 13, 1997, http://www.nytimes.com/books/97/04/13/reviews/970413.13castlet.html.
62. Johann Wolfgang von Goethe, *The Sorrows of Young Werther* (Mineola, NY: Dover Publications, 2002).

63. Isaiah Berlin, *The Roots of Romanticism* (recorded 1965) (London: Chatto & Windus: 1999).
64. Gordon Campbell, *The Hermit in the Garden: From Imperial Rome to Ornamental Gnome* (Oxford: Oxford University Press, 2013).
65. "Jokes by Steven Wright," http://www.wright-house.com/steven-wright/steven-wright-Kn.html, accessed February 28, 2014.
66. Tim Blanning, *The Romantic Revolution: A History (New York:* Random House, 2011).
67. Sophie Fontanel, *The Art of Sleeping Alone: Why One French Woman Suddenly Gave Up Sex* (New York: Scribner, 2011).
68. A.W. Price, *Love and Friendship in Plato and Aristotle* (Oxford: Clarendon Press, 1989).
69. Sigmund Freud, *Three Essays on the Theory of Sexuality*, paperback edition (Eastford, CT: Martino Fine Books, 2011).
70. Paul Bloom, "The Pleasures of Imagination," *Chronicle of Higher Education*, May 30, 2010, http://chronicle.com/article/The-Pleasures-of-Imagination/65678.
71. Paul Bloom, *How Pleasure Works: The New Science of Why We Like What We Like* (New York: W. W. Norton & Company, 2010).

CHAPTER 2: MEET THE BUSINESS ROMANTICS

1. "Hickies.com," http://www.hickies.com/blogs/news/12504485-hickies-announces-key-management-additions-raises-4-2mm-investment-round-to-fuel-global-expansion-and-product-development, accessed February 27, 2014.
2. Alexa Clay, "The Rise of the Intrapreneurs," *Fast Company*, October 3, 2012, http://www.fastcoexist.com/1680655/the-rise-of-the-intrapreneurs.
3. Just as she has helped hundreds of other leaders and groups to create individualized manifestos, Priya also ran a Visioning Lab for me, the original seed of this very book.
4. Peter Sloterdijk, "Critique of Cynical Reason," *Theory and History of Literature* 40 (February 1, 1988).
5. Esther Eidinow and Rafael Ramírez, "'The Eye Of The Soul': Phronesis And The Aesthetics Of Organizing," *Organizational Aesthetics*, 2012, 1 (1), pp. 26-43.

CHAPTER 3: FIND THE BIG IN THE SMALL

1. "WhatIS.com," "Definition Director of First Impressions," http://whatis.techtarget.com/definition/Director-of-First-Impressions, accessed February 28, 2014.
2. Alex Bryson and George MacKerron, "Are You Happy While You

Work?" Center for Economic Performance, Discussion Paper 1187, February 2013, http://cep.lse.ac.uk/pubs/download/dp1187.pdf.

3. Kenneth Matos, "HP & Yahoo's Telecommuting Breakup: It's Not You, It's Me!" *The Huffington Post*, October 23, 2013, http://www.huffingtonpost.com/kenneth-matos/hp-telecommuting-policy_b_4149658.html.
4. Tim Leberecht, "The Office Is Everywhere," *Means the World*, http://meanstheworld.co/work/the-office-is-everywhere.
5. Erwin van der Koogh, "Case-study: Github," *Business in the 21st Century*, http://businessin21stcentury.com/articles/profile-github/, accessed April 11, 2014.
6. It should be mentioned that by the time this book was edited GitHub was embroiled in a scandal over allegations of gender-based harassment at the workplace. Co-founder and CEO Tom Preston-Werner resigned after an internal investigation found no evidence of harassment but concluded that he made "mistakes and errors of judgment." (Nellie Bowles, "GitHub Clears President Tom Preston-Werner, But He Resigns Anyway After Harassment Controversy," *re/code*, April 21, 2014, http://recode.net/2014/04/21/github-ceo-tom-preston-werner-resigns/.) I decided to keep the section in the book because I believe that GitHub's work(place) principles are generally innovative and worth mentioning regardless of the recent events.
7. Michael Chui, James Manyika, Jacques Bughin, Richard Dobbs, Charles Roxburgh, Hugo Sarrazin, Geoffrey Sands, and Magdalena Westergren, "The Social Economy: Unlocking Value and Productivity Through Social Technologies," *McKinsey Global Institute*, July 2012, http://www.mckinsey.com/insights/high_tech_telecoms_internet/the_social_economy.
8. Lisa O'Carroll, "Rebekah Brooks: David Cameron Signed Off Texts 'LOL,'" *Guardian*, May 11, 2012, http://www.theguardian.com/media/2012/may/11/rebekah-brooks-david-cameron-texts-lol.
9. John Gottman, *The Science of Trust: Emotional Attunement for Couples* (New York: W. W. Norton & Company, 2011).
10. Miranda July, "WeThinkAlone.com," http://wethinkalone.com/about/, accessed March 5, 2014.
11. "15 Toasts," http://15toasts.com, accessed May 25, 2014.
12. Jaweed Kaleem, "Death Over Dinner Convenes as Hundreds of Americans Coordinate End of Life Discussions Across U.S.," *Huffington Post*, August 18, 2013, http://www.huffingtonpost.com/2013/08/18/death-over-dinner_n_3762653.html.
13. Life Matters Media, "Death Over Dinner? There's an Appetite for It," April 19, 2013, http://www.lifemattersmedia.org/2013/04/death-over-dinner-theres-an-appetite-for-it/.

CHAPTER 4: BE A STRANGER

1. "MLove," http://www.mlove.com, accessed February 26, 2014.
2. "Venture Strategy for Food Genius," Ideo, accessed March 1, 2014, http://www.ideo.com/work/venture-strategy/.
3. "The 'Siemens Artists-in-Residence Program' Launches in U.S.," *Healthy Hearing*, August 16, 2002, http://www.healthyhearing.com/content/news/Assistance/Awareness/5769-The-siemens-artists-in.
4. "The Recology Artist in Residence Program," http://www.recologysf.com/AIR/index.htm, accessed February 26, 2014.
5. Mark Wilson, "Inside Microsoft Research's First Artist-in-Residence Program," *Fast Company*, December 3, 2013, http://www.fastcodesign.com/3022833/inside-microsoft-researchs-first-artist-in-residence-program.
6. Margaret Sullivan, "Questions on Drones, Unanswered Still," *New York Times*, October 13, 2012, http://www.nytimes.com/2012/10/14/public-editor/questions-on-drones-unanswered-still.html.
7. "Corporate Rebels United," http://corporaterebelsunited.com, accessed June 5, 2014.
8. "Rebels at Work," http://www.rebelsatwork.com, accessed June 5, 2014.
9. Peter Delevett, "Volkswagen, Lego and the 'Beer Garage': Big corporations rushing to partner with Silicon Valley startups," *Mercury News*, December 4, 2013, http://www.mercurynews.com/business/ci_24651658/volkswagen-lego-and-beer-garage-big-corporations-rushing.
10. Maureen Morrison, "McDonald's Sets Up Shop in Silicon Valley," *AdAge*, June 4, 2014, http://adage.com/article/digital/mcdonald-s-sets-shop-silicon-valley/293500/.
11. "Brickfilms," http://brickfilms.com, accessed February 26, 2014.
12. Joe Berkowitz, "Everything About These Pictures of a Tiny, Adventurous Lego Photographer Is Awesome," *Fast Company*, February 26, 2014, http://www.fastcocreate.com/3026935/everything-about-these-pictures-of-a-tiny-adventurous-lego-photographer-is-awesome?utm_source=facebook.
13. Simon Barker, "Lego Fan Asks Girlfriend to Marry Him with Custom Lego Gift Sets," October 30, 2009, http://www.simonbarker.com/lego-fan-asks-girlfriend-to-marry-him-with-custom-lego-gift-sets/.
14. Jeannie Choe, "Trash Talk with Frog's Ashley Menger," *Core* 77, May 18, 2007, http://www.core77.com/blog/education/trash_talk_with_frog_designs_ashely_menger_6363.asp.
15. Linda Tischler, "At Frog Being Green Isn't Easy, It's Essential," *Fast Company*, November 1, 2007, http://www.fastcompany.com/60862/frog-being-green-isn%E2%80%99t-easy-it%E2%80%99s-essential.
16. "Dove Real Beauty Sketches," *YouTube*, https://www.youtube.com/watch?v=litXW91UauE, accessed February 27, 2014.

17. Jakob von Uexküll, *Biological Theory*, translated by Doris L. Mackinnon (New York: Harcourt, Brace & Company, 1926).
18. Janet Cardiff and George Bures Miller, "Walks," http://www.cardiffmiller.com/artworks/walks/index.html, accessed February 26, 2014.
19. Guy Debord, "Introduction to a Critique of Urban Geography," *Situationist International Online*, September 1955, http://www.cddc.vt.edu/sionline/presitu/geography.html.
20. Jon Gertner, "True Innovation," *New York Times*, February 25, 2012, http://www.nytimes.com/2012/02/26/opinion/sunday/innovation-and-the-bell-labs-miracle.html?pagewanted=all&_r=0.
21. Rachel Emma Silverman, "The Science of Serendipity in the Workplace," *Wall Street Journal*, April 30, 2013, http://online.wsj.com/news/articles/SB10001424127887323798104578455081218505870.
22. Business2Community, "Zappos' 11 Company Culture Aspects That Win Over Millennials," *Real Business*, January 17, 2014, http://www.realbusiness.com/2014/01/zappos-11-company-culture-aspects-that-win-over-millennials/.
23. Wendy Kaufman, "Check Out These Gorgeous, Futuristic Tech Company Headquarters," *NPR*, December 24, 2013, http://www.npr.org/blogs/alltechconsidered/2013/12/24/255859905/check-out-these-gorgeous-futuristic-tech-company-headquarters.
24. Eli Pariser, *The Filter Bubble: What the Internet Is Hiding from You* (New York: Penguin Press, 2011).
25. Nicola Clark, "Selecting a Seatmate to Make Skies Friendlier," *New York Times*, February 23, 2012, http://www.nytimes.com/2012/02/24/business/global/selecting-a-seatmate-to-make-skies-friendlier.html?pagewanted=all.
26. "20 Day Stranger," http://www.20daystranger.com, accessed May 26, 2014.
27. And also controversial: Airbnb has been accused of commodifying hospitality, threatening the traditional hotel business, or in some instances even jeopardizing the social fabric of neighborhoods (there have been reports about tensions arising from neighborhoods turning into popular Airbnb tourist destinations, such as Echo Park in Los Angeles).
28. "Carpooling.com," http://www.carpooling.com, accessed February 26, 2014.
29. Nicholas Epley, "Let's make some Metra noise," *Chicago Tribune*, June 3, 2011, http://articles.chicagotribune.com/2011-06-03/opinion/ct-perspec-0605-metra-20110603_1_commuters-quiet-cars-metra-reports.
30. Elizabeth Dunn and Michael Norton, "Hello, Stranger," *New York Times*, April 26, 2014, http://www.nytimes.com/2014/04/26/opinion/sunday/hello-stranger.html.

31. Georg Simmel: *On Individuality and Social Forms*, "The Stranger," 143–50, Donald N. Levine, ed. (Chicago: University of Chicago Press, 1971).
32. Steven Pinker, *The Better Angels of Our Nature: Why Violence Has Declined* (New York: Penguin Press, 2012).

CHAPTER 5: GIVE MORE THAN YOU TAKE

1. "Making Christmas: The View from the Tom and Jerry Christmas Tree," IMDb, http://www.imdb.com/title/tt1881003/, accessed February 27, 2014.
2. "Lights of the Valley," http://lightsofthevalley.com/articles/3650-21st-tree-history.pdf, accessed March 1, 2014.
3. Lewis Hyde, *The Gift: Imagination and the Erotic Life of Property* (New York: Random House, 1979).
4. George Dearing, "Tim O'Reilly: 'Create More Value Than You Capture,'" *Contently*, April 6, 2012, http://contently.com/strategist/2012/04/06/tim-oreilly-value-creation/.
5. Adam Grant, *Give and Take: A Revolutionary Approach to Success* (New York: Viking, 2013).
6. "Frog at SXSW," frog design, http://sxsw.frogdesign.com, accessed February 27, 2014.
7. Sarah Kliff, "Google Nose Is Fake. The Artificial Nose Isn't," *Washington Post*, April 1, 2013, http://www.washingtonpost.com/blogs/wonkblog/wp/2013/04/01/google-nose-is-fake-the-artificial-nose-isnt/s.
8. Other famous brand stunts include Twitter's launch of "Twttr," a free service in which users can only tweet in consonants (for the sake of "denser communications," as the company said); Southwest Airlines' announcement of hot-air balloon travel; or Taco Bell's claim to buy the Liberty Bell (which sparked national outrage).
9. Sylvia Poggioli, "EU Embraces 'Suspended Coffee': Pay It Forward with a Cup of Joe," *NPR*, April 25, 2013, http://www.npr.org/blogs/thesalt/2013/04/24/178829301/eu-embraces-suspended-coffee-pay-it-forward-with-a-cup-of-joe.
10. "Suspended Coffee," Snopes.com, http://www.snopes.com/glurge/suspended.asp, accessed May 25, 2014.
11. "Random Acts of Pizza," http://randomactsofpizza.com, accessed May 3, 2014.
12. "At The Generous Store Chocolates Cost Good Deeds Instead of Money," *Oddity Central*, March 30, 2012, http://www.odditycentral.com/pics/at-the-generous-store-chocolates-cost-good-deeds-instead-of-money.html.
13. Kristin Purcell, "Online Video 2013," Pew Research Center, October 10, 2013, http://pewinternet.org/~/media/Files/Reports/2013/PIP_Online%20Video%202013.pdf.

14. "The Force: Volkswagen Commercial," *YouTube*, https://www.youtube.com/watch?v=R55e-uHQna0, accessed June 4, 2014.
15. "LOL Cats," *YouTube*, https://www.youtube.com/watch?v=RcVyl9X3gFo, accessed June 4, 2014.
16. "Hitler Downfall parodies: 25 worth watching," *Telegraph*, October 6, 2009, http://www.telegraph.co.uk/technology/news/6262709/Hitler-Downfall-parodies-25-worth-watching.html, accessed June 4, 2014.
17. Alison Vingiano, "This Is How a Woman's Offensive Tweet Became the World's Top Story," *Buzzfeed*, December 21, 2013, http://www.buzzfeed.com/alisonvingiano/this-is-how-a-womans-offensive-tweet-became-the-worlds-top-s.
18. Megan Garber, "Batkid: A Heartwarming, Very 2013 Story," *The Atlantic*, November 15, 2013, http://www.theatlantic.com/technology/archive/2013/11/batkid-a-heartwarming-very-2013-story/281560/.
19. Andrew Lasane, "That Adorable 'First Kiss' Video That Everyone Is Talking About Is a Fake," *Complex*, http://www.complex.com/art-design/2014/03/that-first-kiss-video-that-everyone-is-talking-about-is-fake.
20. John Koblin, "A Kiss Is Just a Kiss, Unless It's an Ad for a Clothing Company," *New York Times*, March 14, 2014, http://www.nytimes.com/2014/03/14/business/media/a-kiss-is-just-a-kiss-unless-its-an-ad-for-a-clothing-company.html.
21. NPR Staff, "Scott Simon on Sharing His Mother's Final Moments on Twitter," *NPR*, July 30, 2013, http://www.npr.org/blogs/alltechconsidered/2013/07/30/206987575/Scott-Simon-On-Sharing-His-Mothers-Final-Moments-On-Twitter.
22. Paul Ford and Matt Buchanan, "Death in a Crowd," *New Yorker*, February 4, 2014, http://www.newyorker.com/online/blogs/elements/2014/02/how-social-media-wrote-its-eulogy-for-philip-seymour-hoffman.html?utm_source=www&utm_medium=tw&utm_campaign=20140204.

CHAPTER 6: SUFFER (A LITTLE)

1. Tony Emerson, "The Ten Commandments of IKEA Furniture: Part 1," *Sparefoot Blog*, August 8, 2013, http://blog.sparefoot.com/3928-the-ten-ikea-commandments/.
2. "Ikea or Death," http://ikeaordeath.com, accessed February 27, 2014.
3. Michael I. Norton, Daniel Mochon, and Dan Ariely, "The IKEA-Effect: When Labor Leads to Love," *Journal of Consumer Psychology* 22:3 (July 2012), 453–60.
4. David M. Buss, "The Evolution of Happiness," *American Psychologist* 55 (January 2000), 15–23.
5. "Romantimatic," http://romantimatic.com, accessed February 26, 2014.

6. Chris Matyszczyk, "Progress! An App That Sends a Breakup Text for You," *CNET*, July 27, 2013, http://news.cnet.com/8301-17852_3-57595813-71/progress-an-app-that-sends-a-breakup-text-for-you/.
7. Tony Castle, "Now You Can File For Divorce Online With Wevorce, The H&R Block Of Nasty Breakups," *Fast Company*, April 17, 2014, http://www.fastcoexist.com/3028877/change-generation/now-you-can-file-for-divorce-online-with-wevorce-the-hr-block-of-nasty-bre.
8. It should be noted that this procrastination may have to do with our tendency to separate present and future selves. In a fascinating essay, Alisa Opar contends that we procrastinate because we see our future selves as strangers. She cites a study conducted by Hal Hershfield, an assistant professor at New York University's Stern School of Business, and colleagues, who researched how brain activity changes when people imagine their future and compare it with their present. They came to the conclusion that on a psychological and emotional level we treat our future self as if it's another person. This has real implications for how we make decisions. When we procrastinate, we disassociate our present selves from the consequences of our inaction and let a future version of our self deal with the tasks or problems at hand. We rely on the kindness of strangers, even if that stranger is us. Alisa Opar, "Why We Procrastinate," *Nautilus*, January 16, 2014, http://nautil.us/issue/9/time/why-we-procrastinate.
9. Paul Hayward, "Barcelona's Sense of Style Restores Glory to Blanchflower's Game," *Guardian*, May 30, 2009, http://www.theguardian.com/football/blog/2009/may/31/barcelona-manchester-united-champions-league-glory.
10. "Philosophyfootball.com," http://www.philosophyfootball.com/quotations.php, accessed March 1, 2014.
11. "Corinthians," Fifa.com, http://www.fifa.com/classicfootball/clubs/club=239/, accessed February 27, 2014.

CHAPTER 7: FAKE IT!

1. "*The Great Gatsby* (2013)," *Rotten Tomatoes*, http://www.rottentomatoes.com/m/the_great_gatsby_2013/, accessed February 26, 2014.
2. Simon Hattenstone, "Something to Spray," *Guardian*, July 16, 2003, http://www.theguardian.com/artanddesign/2003/jul/17/art.artsfeatures.
3. Even the fake gets faked: Banksy's success, for example, led to the viral rise of Hanksy, a New York–based street artist who is a mash-up of Banksy and Tom Hanks. This trickster made a name for himself by creating a series of visual gags in secret collaborations with local artists, cleverly augmented by Twitter posts. John Leland, "A Parodist Who Calls Himself Hanksy," *New York Times*, February 14, 2014, http://www.nytimes.com/2014/02/16/nyregion/a-parodist-who-calls-himself-hanksy.html?_r=0.

4. Marc van Gurp, "Unicef: Be a Mom for a Moment," *Osocio*, April 11, 2009, http://osocio.org/message/unicef_be_a_mom_for_a_moment/.
5. Donna Tartt, *The Goldfinch* (New York: Little, Brown and Company, 2013).
6. @NeinQuarterly is a master of irony. Unlike cynicism, irony provides a constructive distance, a persona that allows us to express opinions with nuance and ambiguity. Cynicism celebrates defeat while irony celebrates the plausibility of alternative meanings and interpretations. Romantics appreciate it as an extension of their repertoire to inquire and shed light on the complexity of the world.
7. "Let's Go: Shell in the Arctic," http://arcticready.com, accessed February 27, 2014.
8. Alana Horowitz, "Burger King Twitter Account Hacked," *Huffington Post*, February 18, 2013, http://www.huffingtonpost.com/2013/02/18/burger-king-twitter-hacked_n_2711661.html.
9. Chipotle, "The Scarecrow," *YouTube*, https://www.youtube.com/watch?v=lUtnas5ScSE, accessed February 27, 2014.
10. "Honest Scarecrow," Funnyordie.com, http://www.funnyordie.com/videos/da66b8f1aa/honest-scarecrow, accessed June 5, 2014.
11. Matthew B. Crawford, *Shop Class as Soulcraft: An Inquiry into the Value of Work* (New York: Penguin Press, 2009).
12. Stowe Boyd, "Somewhere Is LinkedIn for the New Way of Work," *GigaOm Research*, January 24, 2014, http://research.gigaom.com/2014/01/somewhere-is-linkedin-for-the-new-way-of-work/.
13. Peter Fulda, "Leaders, Drop Your Masks," *Harvard Business Review*, October 7, 2013, http://blogs.hbr.org/2013/10/leaders-drop-your-masks/.
14. James Surowiecki, "Do the Hustle," *New Yorker*, January 13, 2014, http://www.newyorker.com/talk/financial/2014/01/13/140113ta_talk_surowiecki.
15. Adam Phillips, *On Balance* (New York: Farrar, Straus and Giroux, 2010).
16. Benedict Carey, "Feel Like a Fraud? At Times Maybe You Should," *New York Times*, February 5, 2008, http://www.nytimes.com/2008/02/05/health/05mind.html?_r=0.
17. Amy Cuddy, "Your Body Language Shapes Who You Are," *TED.com*, June 2012, http://www.ted.com/talks/amy_cuddy_your_body_language_shapes_who_you_are.html.
18. Joshua John, "Amy Cuddy: Your Body Language Shapes Who You Are," *MBA@UNC Blog*, July 25, 2013, http://onlinemba.unc.edu/blog/amy-cuddy-your-body-language-shapes-who-you-are-2/.

CHAPTER 8: KEEP THE MYSTIQUE

1. Kingsley Amis, *The James Bond Dossier* (New York: New American Library, 1965).

2. Cima On Mon, "What's a Brand Really Worth?," *Financial Management*, March 18, 2013, http://www.fm-magazine.com/feature/depth/what%E2%80%99s-brand-really-worth.
3. Joseph S. Nye Jr., *Soft Power: The Means to Succeed in World Politics* (New York: PublicAffairs, 2004).
4. Christopher Locke, Rick Levine, Doc Searls, and David Weinberger, *The Cluetrain Manifesto*, http://www.cluetrain.com, accessed March 1, 2014.
5. David Armano, "It's the Conversation Economy, Stupid," *Bloomberg BusinessWeek*, April 9, 2007, http://www.businessweek.com/stories/2007-04-09/its-the-conversation-economy-stupidbusinessweek-business-news-stock-market-and-financial-advice.
6. Baba Shetty and Jerry Wind, "Advertisers Should Act More Like Newsrooms," *Harvard Business Review*, February 15, 2013, http://blogs.hbr.org/2013/02/advertisers-need-to-act-more-like-newsrooms/.
7. Jeff Jarvis, "My Dell Hell," *Guardian*, August 28, 2005, http://www.theguardian.com/technology/2005/aug/29/mondaymediasection.blogging.
8. "United [Airlines] Breaks Guitars," *YouTube*, https://www.youtube.com/watch?v=5YGc4zOqozo, accessed February 27, 2014.
9. Recent examples include backlashes against the new Gap logo; pasta maker SpaghettiOs' "Remember Pearl Harbor with us" tweet showing its mascot holding an American flag; and an allegedly racist tweet by Home Depot.
10. Dawn Kopecki, "JP Morgan's #AskJPM Twitter Hashtag Backfires Against Bank," *Bloomberg*, November 14, 2013, http://www.bloomberg.com/news/2013-11-14/jpmorgan-twitter-hashtag-trends-against-bank.html.
11. Susan Adams, "Don't Fire An Employee And Leave Them In Charge Of The Corporate Twitter Account," *Forbes*, February 1, 2013, http://www.forbes.com/sites/susanadams/2013/02/01/dont-fire-an-employee-and-leave-them-in-charge-of-the-corporate-twitter-account/.
12. Katie Hryce, "Nokia Made A Tweet Swear," *Oyster Magazine*, November 27, 2013, http://www.oystermag.com/nokia-made-a-tweet-swear.
13. John Huey, "How McKinsey Does It," *Fortune*, November 1, 1993, http://money.cnn.com/magazines/fortune/fortune_archive/1993/11/01/78550/.
14. Andrew Ross Sorkin, "McKinsey & Co. Isn't All Roses in a New Book," *New York Times*, September 2, 2013, http://dealbook.nytimes.com/2013/09/02/in-a-new-book-mckinsey-co-isnt-all-roses/?_r=0.
15. Clayton M. Christensen, Dina Wang, and Derek van Bever, "Consulting on the Cusp of Disruption," *Harvard Business Review*, October 2013, http://hbr.org/2013/10/consulting-on-the-cusp-of-disruption/.
16. Maryann Keller, *Rude Awakening: The Rise, Fall and Struggle for Recovery of General Motors* (New York: HarperCollins, 1990).

17. Duff McDonald, *The Firm: The Story of McKinsey and its Secret Influence on American Business* (New York: Simon & Schuster, 2013).
18. Janet Malcolm, "Nobody's Looking at You," *New Yorker*, September 23, 2013, http://www.newyorker.com/reporting/2013/09/23/130923fa_fact_malcolm.
19. Adam Richardson, *Innovation X: Why a Company's Toughest Problems Are Its Greatest Advantage* (San Francisco: Jossey-Bass, 2010).
20. "House of Genius," http://houseofgenius.org, accessed May 24, 2014.
21. Schumpeter, "Anonymous Social Networking. Secrets and Lies," *Economist*, March 22, 2014, http://www.economist.com/blogs/schumpeter/2014/03/anonymous-social-networking.
22. Louise Jury, "Secret Cinema to Kick Off Season of New Films with The Grand Budapest Hotel," *London Evening Standard*, February 17, 2014, http://www.standard.co.uk/goingout/film/secret-cinema-to-kick-off-season-of-new-films-with-the-grand-budapest-hotel-9133327.html.
23. Hallie Sekoff, "13P Implodes: A Decade After Its Debut, the Innovative Playwright Collective Dissolves," *Huffington Post*, September 14, 2012, http://www.huffingtonpost.com/2012/09/13/13ps-implosion_n_1881214.html.
24. Karl Johnson, "Karl Johnson Silhouette Event in Santa Monica," Facebook, May 1, 2011, https://www.facebook.com/events/208860315800264/, accessed February 27, 2014.
25. Lauren Milligan, "Margiela Sans Margiela," *Vogue*, October 3, 2009, http://www.vogue.co.uk/news/2009/10/03/martin-margiela-no-longer-at-the-maison.
26. Eve Oxberry, "Martin Margiela Exits Margiela," *Drapers*, December 9, 2009, http://www.drapersonline.com/news/womenswear/news/martin-margiela-exits-margiela/5008679.article#.UzdFlvldWyU.
27. Patrick Knowles, "Kenya Hara," *SOMA Magazine*, November 2010, http://www.somamagazine.com/kenya-hara/.
28. Some examples: Clorox and S. C. Johnson reveal the ingredients of their consumer products; the start-up Buffer publishes all of its employee salaries; ethics consultancy LRN shares performance reviews—including that of founder and CEO Dov Seidman—with all staff; IDEO open-sources its creative process via OpenIDEO; Internet browser Mozilla turns its strategy development into an open platform.
29. Deyan Sudjic, "The Messy Vitality of Opacity," *Design Mind*, Issue 17, 2012, http://designmind.frogdesign.com/articles/radical-openness/the-messy-vitality-of-opacity.html.

CHAPTER 9: BREAK UP

1. Marina Abramović, Anna Daneri, Giacinto Di Pietrantonio, Lóránd

Hegyi, Societas Raffaello Sanzio, and Angela Vettese, *Marina Abramović* (New York: Charta, 2002); more than thirty years later the two met again, during the MoMA retrospective *The Artist Is Present.* As part of the exhibition, Abramović sat silently at a table and allowed visitors to sit across from her. On the opening day, Ulay did just that, resulting in an emotional reunion.

2. Maria Popova, "Susan Sontag on Marriage," *Brain Pickings*, August 17, 2012, http://www.brainpickings.org/index.php/2012/08/17/susan-sontag-on-marriage/.
3. Miyoung Kim, "Samsung Sets New Smartphone Sales Record in Fourth Quarter, Widens Lead Over Apple: Report," *Reuters*, January 28, 2014, http://www.reuters.com/article/2014/01/28/us-samsung-sales-idUSBREA0R00S20140128.
4. "Atari graveyard found: Millions of E.T. cartridges, legendary 'worst video game ever'," *Associated Press*, April 26, 2014, http://www.mercurynews.com/ci_25644581/hundreds-atari-e-t-games-surprise-landfill-diggers.
5. Andrew Mason, "Groupon Founder Andrew Mason's Farewell Letter to Employees—full text," *Guardian*, March 1, 2013, http://www.theguardian.com/technology/blog/2013/mar/01/groupon-andrew-mason-fired-letter.
6. Jill Ahlberg Yohe, "Situated Flow: A Few Thoughts on Reweaving Meaning in the Navajo Spirit Pathway," *Museum Anthropology Review* 6:1 (2012), https://scholarworks.iu.edu/journals/index.php/mar/article/view/1033/2037.

CHAPTER 10: SAIL THE OCEAN

1. Joan Didion, *The White Album* (New York: Farrar, Straus and Giroux, 1979).
2. Heidi Grant Halvorson, "How Happiness Changes with Age," *Atlantic,* May 28, 2013, http://www.theatlantic.com/health/archive/2013/05/how-happiness-changes-with-age/276274/.
3. David Finegold, Susan Mohrman, and Gretchen M. Spreitzer, "Age Effects on the Predictors of Technical Workers' Commitment and Willingness to Turnover," *Journal of Organizational Behavior* 23:5 (August 2002), 655–74.
4. Dennis Lim, "Nine More Years On, and Still Talking," *New York Times,* May 3, 2013, http://www.nytimes.com/2013/05/05/movies/ethan-hawke-and-julie-delpy-discuss-before-midnight.html?_r=0.
5. "Declining Employee Loyalty: A Casualty of the New Workplace," *Knowledge @ Wharton*, May 09, 2012, https://knowledge.wharton.upenn.edu/article/declining-employee-loyalty-a-casualty-of-the-new-workplace, accessed May 31, 2014.
6. "Brian Eno by Alfred Dunhill," *YouTube*, https://www.youtube.com/watch?v=5mqtc2Z3K8o&feature, accessed March 26, 2014.

7. "Cultural Capital," http://en.wikipedia.org/wiki/Cultural_capital, accessed March 1, 2014.
8. Jonathan Franzen, "Liking Is for Cowards. Go for What Hurts," *New York Times*, May 29, 2011, http://www.nytimes.com/2011/05/29/opinion/29franzen.html?pagewanted=all&_r=0.
9. Andrew Bird, "Puzzling Through (A Love Song), *New York Times*, August 30, 2013, http://opinionator.blogs.nytimes.com/2013/08/30/puzzling-through-a-love-song/?_php=true&_type=blogs&_r=0.
10. John Cage, *Silence: Lectures and Writings, 50th Anniversary Edition* (Middletown, CT: Wesleyan University Press, 2013).

CHAPTER 11: TAKE THE LONG WAY HOME

1. Katie Roiphe, "The Allure of Messy Lives," *New York Times*, July 30, 2010, http://www.nytimes.com/2010/08/01/fashion/01Cultural.html?pagewanted=all&_r=0.
2. Ibid.
3. Alexa Clay, "The Amish Futurist and the power of buttermilk," *re:publica 14*, May 8, 2014, http://re-publica.de/en/session/amish-futurist-and-power-buttermilk.
4. Charles Yu, "Happiness Is a Warm iPhone," *New York Times*, February 22, 2014, http://www.nytimes.com/2014/02/23/opinion/sunday/happiness-is-a-warm-iphone.html?_r=0.
5. "School of Poetic Computation," http://sfpc.io, accessed May 26, 2014.
6. Tim Leberecht, "Autonomy Is Essential for Innovation," *Means the World*, http://meanstheworld.co/work/autonomy-essential-innovation.
7. "Nostalgia," University of Southampton, http://www.southampton.ac.uk/nostalgia/publications, accessed March 14, 2014.
8. "Bond Gifts," http://bondgifts.com, accessed June 8, 2014.
9. "Think Clearly," http://www.thnkclrly.com/newsletter, accessed May 26, 2014.
10. "The Rumpus: Letters In The Mail," http://therumpus.net/letters, accessed May 26, 2014.
11. "Cowbird," http://cowbird.com, accessed May 26, 2014.
12. Kristina Loring, "The Never-Ending Story," *Design Mind*, Issue 14, 2011, http://designmind.frogdesign.com/articles/the-never-ending-story.html.
13. "FutureMe," http://futureme.org, accessed May 26, 2014.
14. Stuart Elliott, "New Bow for CARE Packages," *New York Times*, November 6, 2013, http://www.nytimes.com/2013/11/07/business/media/new-bow-for-care-packages.html?_r=0.
15. "Brain Pickings," http://www.brainpickings.org, accessed May 26, 2014.
16. Orham Pamuk, *The Museum of Innocence* (New York: Knopf, 2009).
17. "Voyager – The Interstellar Mission," NASA, Jet Propulsion Laboratory,

California Institute of Technology, http://voyager.jpl.nasa.gov/spacecraft/goldenrec.html, accessed May 26, 2014.

CHAPTER 12: STAND ALONE, STAND BY, STAND STILL

1. "Louis C.K. Hates Cell Phones," *TeamCoco*, http://teamcoco.com/video/louis-ck-springsteen-cell-phone, accessed February 27, 2014.
2. Barbara L. Fredrickson, "Your Phone vs. Your Heart," *New York Times*, March 23, 2013, http://www.nytimes.com/2013/03/24/opinion/sunday/your-phone-vs-your-heart.html?_r=0.
3. Alvin Toffler, *Future Shock* (New York: Random House, 1970).
4. Douglas Rushkoff, *Present Shock: When Everything Happens Now* (New York: Current, 2013.
5. Linda Stone, "Continuous Partial Attention," *lindastone.net*, http://lindastone.net/qa/continuous-partial-attention/, accessed February 26, 2014.
6. Anand Giridharadas, "Exploring New York, Unplugged and on Foot," *New York Times*, January 24, 2013, http://www.nytimes.com/2013/01/25/nyregion/exploring-red-hook-brooklyn-unplugged-and-with-friends.html?pagewanted=all.
7. Ibid.
8. Gavin Pretor-Pinney, "Live with Your Head in the Clouds," *TED.com*, June 13, 2013, http://blog.ted.com/2013/06/13/live-with-your-head-in-the-clouds-gavin-pretor-pinney-at-tedglobal-2013/.
9. Sarah Whyte, "Ban Mobile Phones, Retailers Say," *Sydney Morning Herald*, July 4, 2013, http://www.smh.com.au/digital-life/mobiles/ban-mobile-phones-retailers-say-20130703-2pbzr.html.
10. "Four Seasons Resort Costa Rica at Peninsula Papagayo Partners with iPhone Case Company Uncommon on New Disconnect to Reconnect Program for Guests," Four Seasons Press Room, November 26, 2012, http://press.fourseasons.com/costarica/hotel-news/2012/11/four_seasons_resort_costa_rica_at_peninsula_papaga/, accessed March 3, 2014.
11. "Lessons in Customer Service from the World's Most Beloved Companies," *Help Scout*, October 9, 2013, https://www.helpscout.net/blog/customer-focus/.
12. Eric Lowitt, *The Collaboration Economy: How to Meet Business, Social, and Environmental Needs and Gain Competitive Advantage* (San Francisco: Jossey-Bass, 2013).
13. John M. Bernard, *Business at the Speed of Now: Fire Up Your People, Thrill Your Customers, and Crush Your Competitors* (New York: Wiley, 2011).
14. Keith Ferrazzi, *Never Eat Alone: And Other Secrets to Success, One Relationship at a Time* (New York: Crown Business, 2005).

15. "Serpentine Galleries," http://www.serpentinegalleries.org/exhibitions-events/experiment-marathon, accessed March 28, 2014.
16. J. Keith Murnighan, *Do Nothing! How to Stop Overmanaging and Become a Great Leader* (New York: Penguin Press, 2012).
17. Schumpeter, "In Praise of Laziness," *Economist,* August 17, 2013, http://www.economist.com/news/business/21583592-businesspeople-would-be-better-if-they-did-less-and-thought-more-praise-laziness.
18. Michael Bar-Eli, Ofer H. Azar, Ilana Ritov, Yael Keidar-Levin, and Galit Schein, "Action Bias Among Elite Soccer Goalkeepers: The Case of Penalty Kicks," *Journal of Economic Psychology*, Elsevier, 28:5 (October 2005), 606–21.
19. William Deresiewicz, "Solitude and Leadership," *American Scholar,* Spring 2010, http://theamericanscholar.org/solitude-and-leadership.

CHAPTER 13: MEASURES OF SUCCESS

1. Joel Lovell, "George Saunders's Advice to Graduates," *New York Times,* July 31, 2013, http://6thfloor.blogs.nytimes.com/2013/07/31/george-saunderss-advice-to-graduates/?_r=0.
2. Morten T. Hansen, "IDEO CEO Tim Brown: T-Shaped Stars: The Backbone of IDEO's Collaborative Culture," *ChiefExecutive.net,* January 21, 2010, http://chiefexecutive.net/ideo-ceo-tim-brown-t-shaped-stars-the-backbone-of-ideoae%E2%84%A2s-collaborative-culture.

CHAPTER 14: THE NEW ROMANTIC AGE

1. Arlie Russell Hochschild, *The Outsourced Self: What Happens When We Pay Others to Live Our Lives for Us* (New York: Metropolitan Books, 2012).
2. Abraham Maslow, "A Theory of Human Motivation," *Psychological Review,* 50, 370–96, 1943. Retrieved from http://psychclassics.yorku.ca/Maslow/motivation.htm.
3. Miya Tokumitsu, "In the Name of Love," *Jacobin Magazine*, Issue 13, January 2014, https://www.jacobinmag.com/2014/01/in-the-name-of-love/.
4. Jan Chipchase, "You Lookin' at Me? Reflections on Google Glass," *All Things D*, April 12, 2013, http://allthingsd.com/20130412/you-lookin-at-me-reflections-on-google-glass/.
5. Richard Marggraf Turley, "Melting the Wax . . . Romantic Surveillance," richardmarggrafturley.weebly.com,http://richardmarggrafturley.weebly.com/1/previous/3.html, accessed January 5, 2014.
6. Jeremy Bentham, *Deontology,* ed. John Bowring (London: Longman, Rees, Orme, Brown, Green, and Longman, 1834).
7. Take, for example, the famous friendship between the poets

Wordsworth and Coleridge. Some of the period's greatest poems in the English language were created when these two men retreated to North Devon to write; they walked in the countryside, sharing ideas and challenging each other, often accompanied by Wordsworth's erudite sister, Dorothy. In fact, the three figures soon became notorious in and around North Devon, and it wasn't long before they aroused suspicion among the locals, who pegged the roving bohemians as revolutionaries sympathetic to France—Public Enemy Number One in Britain at the time. This speculation was only fanned by their association with the political radical John Thelwall, whom government officials considered to be "the most dangerous man in Britain." Eventually, the British Home Office sent a spy to North Devon to gather intelligence on the rebel poets, whose romantic writings were considered an affront to the supremacy of reason and usefulness. In a later account, Coleridge ridiculed the stupidity of the government stooge, noting that he mistook the poets' conversation about Spinoza as an account of "Spy Nozy." Still, despite this mocking, the surveillance was very real. The poets had to disassociate themselves from Thelwall and adopted allegory, ambiguity, and irony as a means of cloaking more subversive discourse. Despite these efforts, William Wordsworth and fellow romantic John Keats were eventually put under state censure by then–prime minister William Pitt.

8. Scott Kirsner, "Wireless Worker Badges Hold Promise—and Problems," *Boston Globe,* November 3, 2013, http://www.bostonglobe.com/business/2013/11/02/breakthrough-management-tool-big-brother-workplace/WKMDFFieBC9M98EWUPbFZL/story.html?s_campaign=sm_tw.
9. Evgeny Morozov, "The Real Privacy Problem," *MIT Technology Review,* October 22, 2013, http://www.technologyreview.com/featuredstory/520426/the-real-privacy-problem/.
10. "Tim O'Reilly Google+ Page," https://plus.google.com/+TimOReilly/posts/CPiAX9YiVUB, accessed February 11, 2014.
11. Graham Greene, *The Heart of the Matter* (New York: Vintage, 2001; original: 1948).
12. John Keats, "Selections from Keats's Letters (1817)," *Poetry Foundation,* http://www.poetryfoundation.org/learning/essay/237836?page=2, accessed February 27, 2014.
13. Lesley Hazleton, "The doubt essential to faith," *TED.com,* June 2013, http://www.ted.com/talks/lesley_hazleton_the_doubt_essential_to_faith.
14. Quentin Hardy, "What's Lost When Everything Is Recorded," *New York Times,* August 17, 2013, http://bits.blogs.nytimes.com/2013/08/17/whats-lost-when-everything-is-recorded/.
15. Carl DiSalvo, *Adversarial Design* (Cambridge: MIT Press, 2012).

THE BUSINESS ROMANTIC STARTER KIT

1. Deviant, "Arms and Sleepers: Nostalgia For The Absolute," *Sputnik Music*, March 11, 2011, http://www.sputnikmusic.com/review/42319/Arms-and-Sleepers Nostalgia-For-The-Absolute/.
2. "Arvo Pärt - Cantus in Memory of Benjamin Britten," *YouTube*, https://www.youtube.com/watch?v=TRZLxxR23K4, accessed June 8, 2014.
3. "Luke Addingon Furniture," http://addingtonfurniture.tumblr.com/post/80710026203, accessed June 1, 2014.
4. Roger Ebert, "Great Movie: *Citizen Kane*," http://www.rogerebert.com/reviews/great-movie-citizen-kane-1941, accessed March 2, 2014.

INDEX

ABOUT THE AUTHOR

Tim Leberecht is the chief marketing officer of NBBJ, a global design and architecture firm. Previously, he was the chief marketing officer of product design and strategy firm Frog Design. His writing has appeared in publications such as *Fast Company*, *Forbes*, *Fortune*, and *Wired*. He has spoken at venues including TEDGlobal, The Economist Big Rethink, DLD, and the Silicon Valley Bank CEO Summit. Leberecht is the founder and former publisher of the award-winning *Design Mind* magazine, the co-founder of the *15 Toasts* dinner series, and an adviser to The Human Agency, a collective of social change–makers. He serves on the World Economic Forum's Global Agenda Council on Values and on the board of Jump Associates, a strategy and innovation consultancy. He lives in San Francisco with his wife and daughter.